AF898728

JANIRE NÁJERA

ATOMIC—ED

PAGE—3
Edward Bernard Grothus
A look through the archive of the legendary Atomic Ed

PAGE—81
CONFIDENTIAL
A selection of Ed's correspondence from 1971 to 2007

PAGE—113
The Black Hole
A photo-essay documenting artefacts found
during the liquidation sale in 2012

PAGE—161
Absent Presences, Unconscious Collaborators
Celia Rose Jackson

PAGE —164
Photo Captions

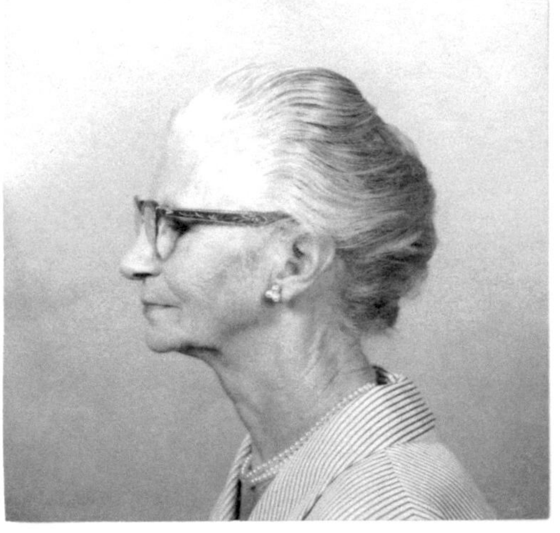

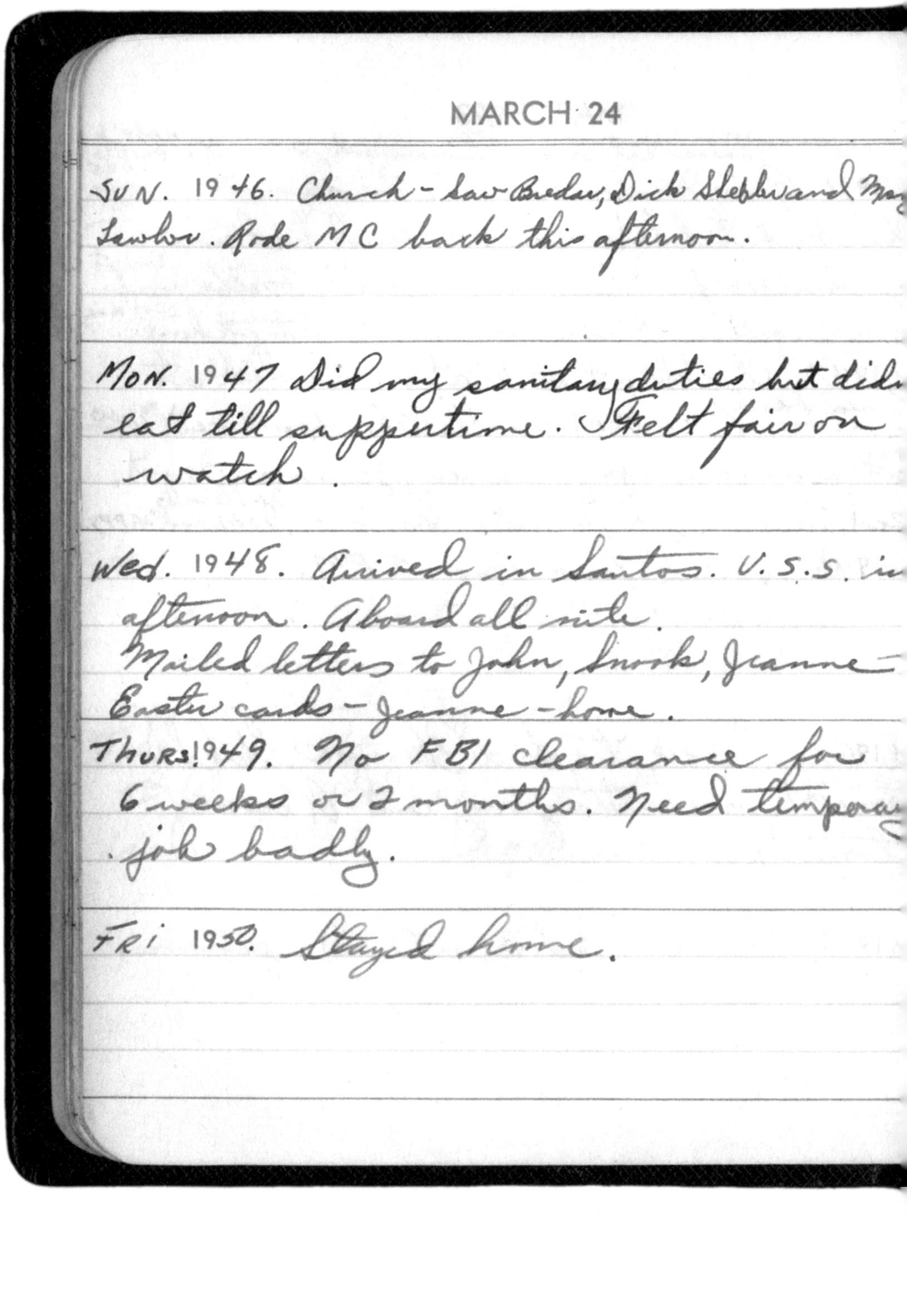

MARCH 24

SUN. 1946. Church - saw Bredar, Dick Shepherd and Mag
Lawlor. Rode MC back this afternoon.

MON. 1947 Did my sanitary duties but didn
eat till suppertime. Felt fair on
watch.

Wed. 1948. Arrived in Santos. U.S.S. in
afternoon. Aboard all nite.
Mailed letters to John, Snook, Jeanne -
Easter cards - Jeanne - home.

THURS. 1949. No FBI clearance for
6 weeks or 2 months. Need temporary
job badly.

FRI 1950. Stayed home.

MARCH 25

1946. Surveyed again. M.C. battery fell out on street and broke to pieces. Can't get another one here. Ordered one from MC shop.

1947 Feel much better today. Eating regular again. Beautiful day. Put up ariel.

1948 Left dock at 3AM but line fouled screw so didn't actually leave the harbor till 1130 AM.

1949. Getting caught up with letter writing. Joe and Ann had Joe Jr. at 5:30 AM at 6½ lbs.

1960. Worked. Went to Ski Carnival Ball at La Fonda. Also saw "the Great Dan Patch" at Lensic. Stopped at El Nido.

UNIVERSITY OF CALIFORNIA
LOS ALAMOS SCIENTIFIC LABORATORY
(CONTRACT, W-7405, ENG. 36)

EMPLOYMENT OFFICE
P. O. BOX 1663
LOS ALAMOS, NEW MEXICO

April 18, 1949

In Reply Refer to:

Mr. Edward B. Grothus
2723 Santa Monica Drive
Albuquerque, New Mexico

Dear Mr. Grothus:

Confirming our recent conversation concerning employment with our Laboratory, we are pleased to offer you a position as a Machinist, Second Class, at a rate of $1.70 per hour, which we sincerely hope you will find acceptable. While Laboratory operations are generally on a standard 40 hour work week, the department to which you will be assigned has been operating on a 44 hour week for some time, with time and one-half for all hours over 8 in one day, or 40 in one week. We cannot, however, guarantee that this 44 hour provision will continue indefinitely, but thought you would be interested inasmuch as it will affect you currently.

Employment at the Laboratory is contingent upon your passing a required physical examination, for which forms and instructions are attached, and upon security clearance. We have requested security clearance, which usually requires 60 to 90 days for completion. Please make no plans for coming to Los Alamos until we notify you that both the security clearance and the physical examination have been approved.

After six months service, an automatic deduction must be made from your salary toward the California State Employees Retirement System, as explained in detail in the attached booklet.

We have reserved a single dormitory room for you with the Los Alamos Housing Office and we shall assist you in every way possible in making satisfactory arrangements. However, because the Housing Office is not operated by the Laboratory, it will be necessary for you, upon your arrival, to discuss possible specific dormitory assignments directly with that office and to complete the arrangements yourself. It should be noted that this in most cases gives you a certain degree of choice with regard to the housing unit assigned, something which would not be possible were all arrangements to be made in advance.

If you find the conditions stated herein acceptable, please sign the attached copy and return it to us at your earliest convenience.

Very truly yours,

E. W. Doty
Ass't. Employment Director

TAD:hj
Enc: Employees' Handbook
Los Alamos Guide
Medical Examination Forms
Arrival Procedure
Rent Schedule
Tax Exemption Certificates
SERS Booklet

UNIVERSITY OF CALIFORNIA
LOS ALAMOS SCIENTIFIC LABORATORY
(CONTRACT, W-7405, ENG. 36)

EMPLOYMENT OFFICE
P. O. BOX 1663
LOS ALAMOS, NEW MEXICO

In Reply Refer to:

May 9, 1949

Mr. Edward B. Grothus
2723 Santa Monica Drive
Albuquerque, New Mexico

Dear Mr. Grothus:

On April 18 we made you an offer of employment as a Machinist, Second Class, at the rate of $1.70 per hour, requesting that you return to us a signed attached copy if you found the terms of the offer acceptable.

To date we have not heard from you in this respect. We will greatly appreciate your letting us have your decision by return mail.

Hoping to hear from you soon, I am

Sincerely yours,

Conrad Latour
Employment Supervisor

CL:ge

UNIVERSITY OF CALIFORNIA
LOS ALAMOS SCIENTIFIC LABORATORY
(CONTRACT, W-7405, ENG. 36)

EMPLOYMENT OFFICE
P. O. BOX 1663
LOS ALAMOS, NEW MEXICO

June 6, 1949

In Reply Refer to:

Mr. Edward B. Grothus
794 A 46th St.
Los Alamos, N.M.

Dear Mr. Grothus:

Thank you for your signed acceptance of our recent letter of offer, and we are looking forward to your joining our staff.

It will be quite in order for you to take your medical examination at Davenport, and we shall advise you just as soon as it is approved. Your security clearance should be forthcoming within the next four weeks.

Please keep me informed of your forwarding address so that I can get in touch with you as soon as your medical and/or security approval has been received.

Yours very truly,

Conrad Latour
Employment Supervisor

CL:lw

IDENTIFICATION CARD
ATOMIC ENERGY
COMMISSION
LOS ALAMOS FIELD OFFICE
LOS ALAMOS, N. M.
NAME Edward B. GROTHUS
EMPLOYER U. of C. R
DATE OF BIRTH
EYES
HEIGHT
6/28/23
Brn
165
Z-6251
al
DIRECTOR, SECURITY DIVISION
PROPERTY OF U S GOVT
26251
IDENTIFICATION CARD

High s
C

unity center
we live here
New Hospital

ONE BOMB is TOO MANY

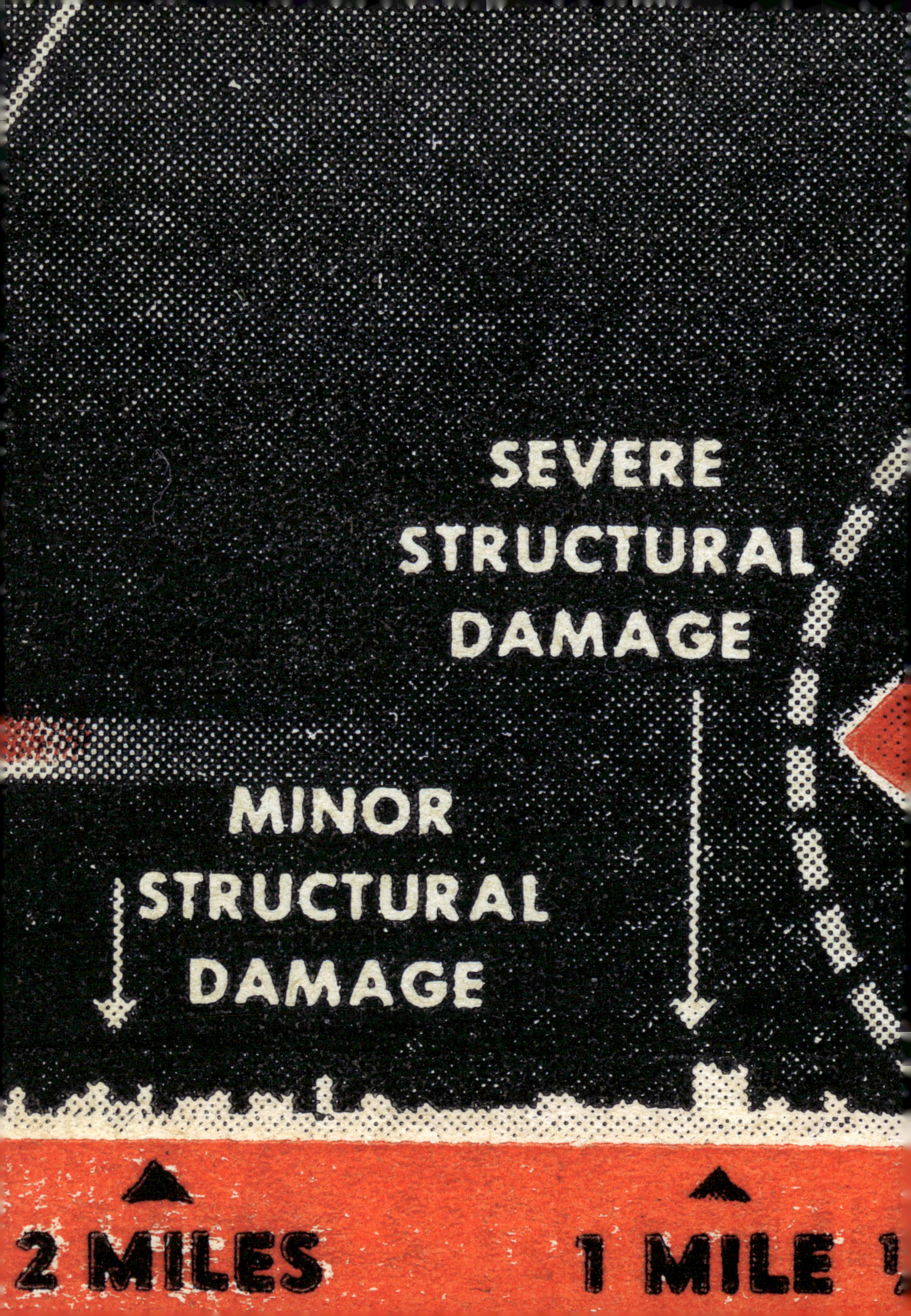
SEVERE
STRUCTURAL
DAMAGE
MINOR
STRUCTURAL
DAMAGE
2 MILES
1 MILE

COMPLETE
DESTRUCTION
MILE 0 ½ MILE 1

western union

Telegram

NO. WDS.—CL. OF SVC.	PD. OR COLL.	CASH NO.	CHARGE TO THE ACCOUNT OF	☐ OVER NIGHT TELEGRAM UNLESS BOX ABOVE IS CHECKED THIS MESSAGE WILL BE SENT AS A TELEGRAM

Send the following message, subject to the terms on back hereof, which are hereby agreed to

4/14/70 19

TO Presiden Nixon
CARE OF OR APT. NO.

STREET & NO. The White House
TELEPHONE

CITY & STATE Washington, DC.
ZIP CODE

Dear Mr. President—
Stop waging war. Stop your stupid war.

Sincerely— Los Alamos Sales
Edward B. Geothu
208 Andanada
Los Alamos, N.M.

2.75

D 4.01

western union

Telegram

NO. WDS.—CL. OF SVC.	PD. OR COLL.	CASH NO.	CHARGE TO THE ACCOUNT OF	☐ OVER NIGHT TELEGRAM UNLESS BOX ABOVE IS CHECKED THIS MESSAGE WILL BE SENT AS A TELEGRAM

Send the following message, subject to the terms on back hereof, which are hereby agreed to

19

TO Mr. Heney H. Fowler
CARE OF OR APT. NO.

STREET & NO. Secretary of the Treasury
TELEPHONE

CITY & STATE Washington, D.C.
ZIP CODE

Dear Mr. Secretary—
How many billions of dollars can this great country of ours spend waging war before we destroy ourselves? Please advise.

Sincerely—
Los Alamos Sales C.
Box 795
Los Alamos N.M.

SENDER'S TEL. NO. 662-7438 [redacted] Edward B. Geothus
Manager

WU 1207 (R 5-69)

western union **Telegram**

O. WDS.–CL. OF SVC.	PD. OR COLL.	CASH NO.	CHARGE TO THE ACCOUNT OF	☐ OVER NIGHT TELEGRAM UNLESS BOX ABOVE IS CHECKED THIS MESSAGE WILL BE SENT AS A TELEGRAM

end the following message, subject to the terms on back hereof, which are hereby agreed to — 3/3/71 19

O: PRESIDENT RICHARD NIXON — CARE OF OR APT. NO.

TREET & NO.: THE WHITE HOUSE — TELEPHONE

TY & STATE: WASHINGTON, D.C. — ZIP CODE

DEAR MR. PRESIDENT:

SHOULD YOU USE ATOMIC WEAPONS IN INDOCHINA KINDLY REMEMBER THAT IT IS OUR FORCES THAT ARE CONCENTRATED AND THEY WILL MAKE A VERY ATTRACTIVE TARGET FOR THE OTHER SIDES ATOMIC WEAPONS. WRITTEN FROM THE ATOMIC CITY.

SINCERELY,

EDWARD B. GROTHUS
BOX 795
LOS ALAMOS, NEW MEXICO

---

western union **Telegram**

O. WDS.–CL. OF SVC.	PD. OR COLL.	CASH NO.	CHARGE TO THE ACCOUNT OF	X OVER NIGHT TELEGRAM UNLESS BOX ABOVE IS CHECKED THIS MESSAGE WILL BE SENT AS A TELEGRAM

end the following message, subject to the terms on back hereof, which are hereby agreed to — March 23, 1971 19

O: President Nixon — CARE OF OR APT. NO.

TREET & NO.: The White House — TELEPHONE

TY & STATE: Washington, D.C. — ZIP CODE

Dear Mr. President:

People do not wage war. Leaders wage war. You are our leader. You wage war all the while crying peace. You are a deceiver and a misleader. You destroy our rescources, our people, our reputation every morning when you dispatch our obedient sons riding in our bombers. Stop giving barbaric commands. Great leaders have a sense of magnanimity. Please show some magnanimity to the people you daily treat so harshly.

Sincerely,

Edward B. Grothus
208 Andanada
Los Alamos, New Mexico 87544

NDER'S TEL. NO. 505-662-2539 — NAME & ADDRESS: Same

U 1207 (R 5-69)

May 10th, 1972

Hi:

Just want to thank you for your faithfullness in sending the check. Also the horror stories of V. Nam which are so true. I'm sorry for any part I have played in it all. With so many objectors I will never understand why more progress isnt being made to avoid the horrors.

Nixon is diabolical----with the mining and other things going on.

Fr. Smith and others here are doing all the others are doing "just protesting"

Nixon is very lucky that he hasn't been killed long before now.

What will the next "ruler" do?

As always,

Mother

western union Telegram

NO. WDS.-CL. OF SVC.	PD. OR COLL.	CASH NO.	CHARGE TO THE ACCOUNT OF	X OVER NIGHT TELEGRAM. UNLESS BOX ABOVE IS CHECKED THIS MESSAGE WILL BE SENT AS A TELEGRAM

Send the following message, subject to the terms on back hereof, which are hereby agreed to

May 5 197

TO President Nixon

CARE OF OR APT. NO.

STREET & NO. The White House

TELEPHONE

CITY & STATE Washington, D.C.

ZIP CODE

Dear Mr. President:

Are you going to bomb in perpetuity? Are you going to destroy each town and village as it falls into Vietnamese hands in order to save it? You are as sane as Adolph Eichman but can't you see that what you are doing is wrong? Do you not know the word magnanimity? It is noble for the mighty to show magnanimity. You have dropped mote bombs and created more destruction than any man in the history of the world. Are you not yet sated with blood? Please stop your idiotic quest to destroy Communism in this way. You must instead build-build-build.

Very sincerely,

Edward B. Grothus
208 Andanada
Los Alamos, New Mexico 87544

Phone 505-662-2539

western union Telegram

NO. WDS.-CL. OF SVC.	PD. OR COLL.	CASH NO.	CHARGE TO THE ACCOUNT OF	OVER NIGHT TELEGRAM. UNLESS BOX ABOVE IS CHECKED THIS MESSAGE WILL BE SENT AS A TELEGRAM

Send the following message, subject to the terms on back hereof, which are hereby agreed to

12/19 19

TO President Nixon

CARE OF OR APT. NO.

STREET & NO. The White House

TELEPHONE

CITY & STATE Washington, D.C.

ZIP CODE

Dear Mr. President –

May no one ever do unto you and your supporters what you do unto other. You are the greatest destroyer alive today.

SENDER'S TEL. NO. 662-2539 NAME & ADDRESS Edward B. Grothus 208 Andanada Los Alamos, New Mexico

WU 1207 (R 5-69)

western union

Telegram

NO. WDS.-CL. OF SVC.	PD. OR COLL.	CASH NO.	CHARGE TO THE ACCOUNT OF	X OVER NIGHT TELEGRAM deliver UNLESS BOX ABOVE IS CHECKED THIS MESSAGE WILL BE SENT AS A TELEGRAM

Send the following message, subject to the terms on back hereof, which are hereby agreed to

July 2 1973

CARE OF OR APT. NO.

TO President Richard Nixon

STREET & NO. The White House

TELEPHONE

CITY & STATE Washington, D.C.

ZIP CODE

Dear Mr. President:

I think it is so very nice that you won't be able to kill, bomb, burn, and destroy beyond August 15th. I hope that you are not too unhappy that the congress has finally put an end to your terrible actions.

Sincerely, Edward B. Grothus

Edward B. Grothus
208 Andanada
Los Alamos, New Mexico 87544

3.00 paid
7/2/73

western union

Telegram

NO. WDS.-CL. OF SVC.	PD. OR COLL.	CASH NO.	CHARGE TO THE ACCOUNT OF	OVER NIGHT TELEGRAM UNLESS BOX ABOVE IS CHECKED THIS MESSAGE WILL BE SENT AS A TELEGRAM

Send the following message, subject to the terms on back hereof, which are hereby agreed to

October 22, 1974 19

CARE OF OR APT. NO.

TO President Gerald Ford

STREET & NO. The White House

TELEPHONE This item is to be delivered.

CITY & STATE Washington, D.C.

ZIP CODE

Dear Mr. President:

One atomic weapon can kill one million people in one second and simultaneously level the city in which they live. I can think of no reason which would be sufficient for me to do such a thing. You now have the power to maintain this deadly process for one hundred and forty minutes. Your military experts and yourself are still unsated and want more and more of these deadly devices. Is there no limit to the energy you expend to maintain this deadly business? I think America needs other things more.

3.56 paid
10/22/74

Sincerely,

Edward B. Grothus
208 Andanada
Los Alamos, New Mexico 87544

SENDER'S TEL. NO. 505-662-2539

NAME & ADDRESS

WU 1207 (R 5-69)

THE WHITE HOUSE

Mr. Ed Grothus
Box 795
Los Alamos, New Mexico

THE WHITE HOUSE

WASHINGTON

April 17, 1975

Dear Mr. Grothus:

On President Ford's behalf, I want to acknowledge your telegram and to assure you that your comments have been noted. Your interest in sharing your views is appreciated.

Sincerely,

Roland L. Elliott

Roland L. Elliott
Director of Correspondence

Mr. Ed Grothus
Box 795
Los Alamos, New Mexico

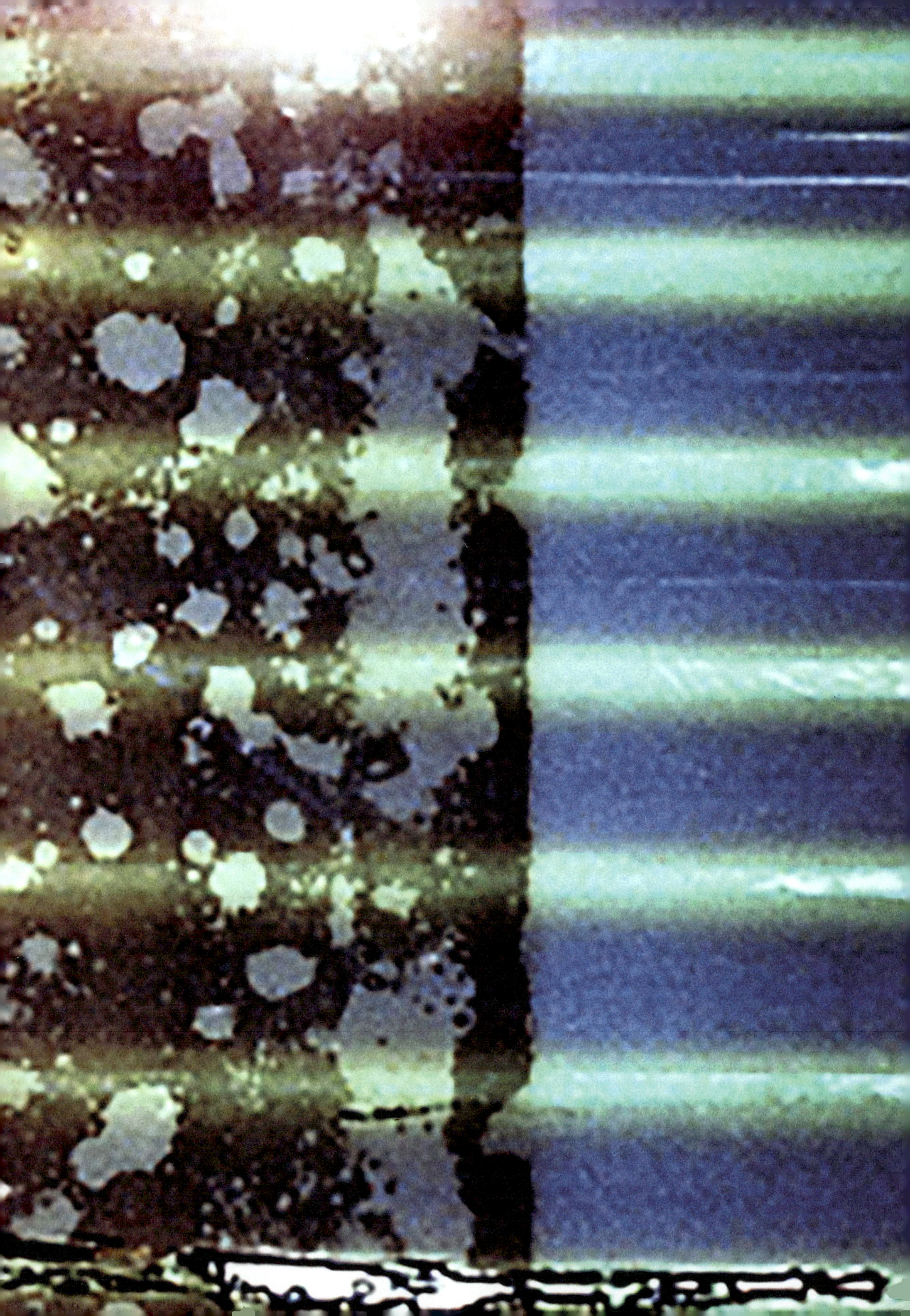

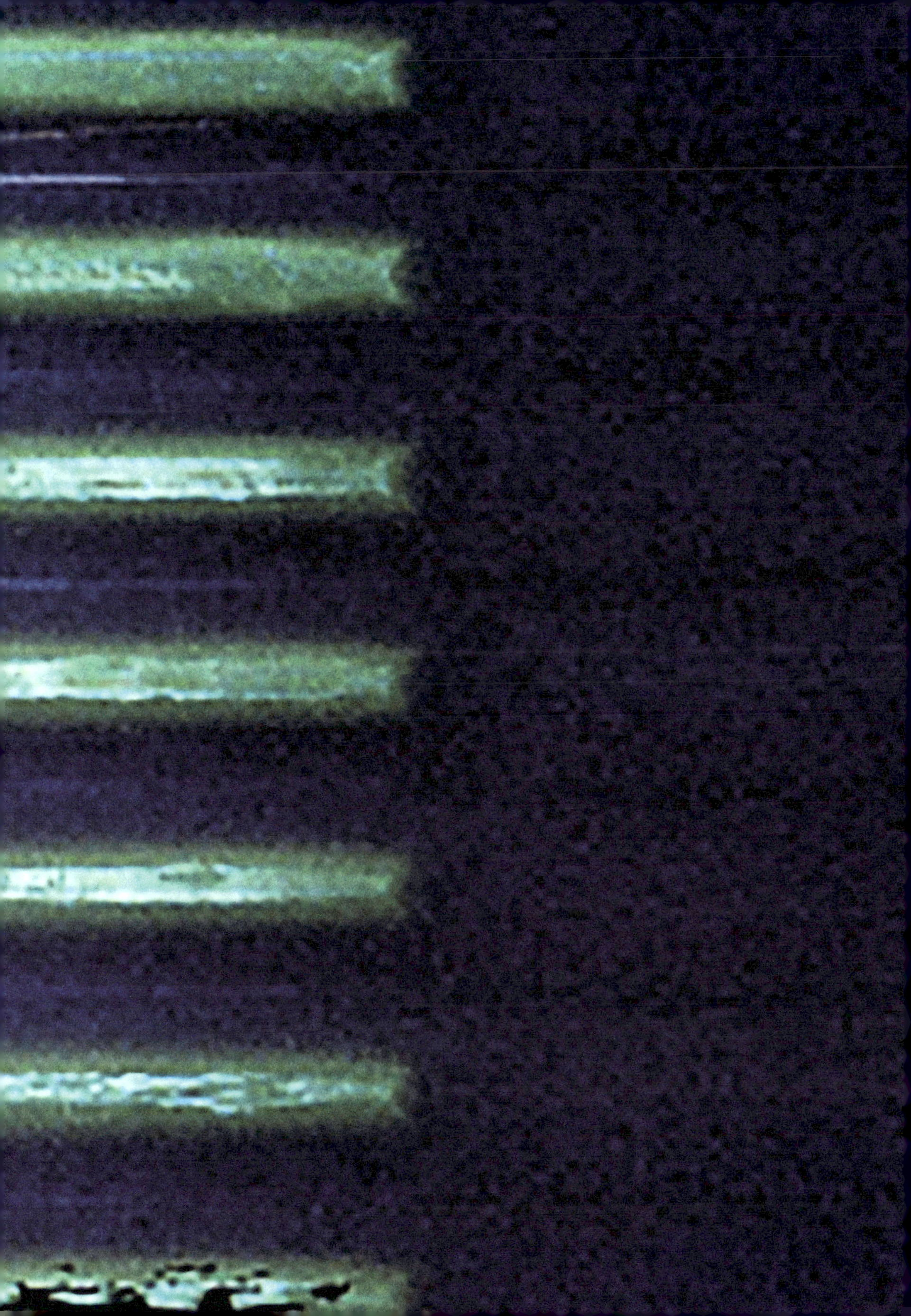

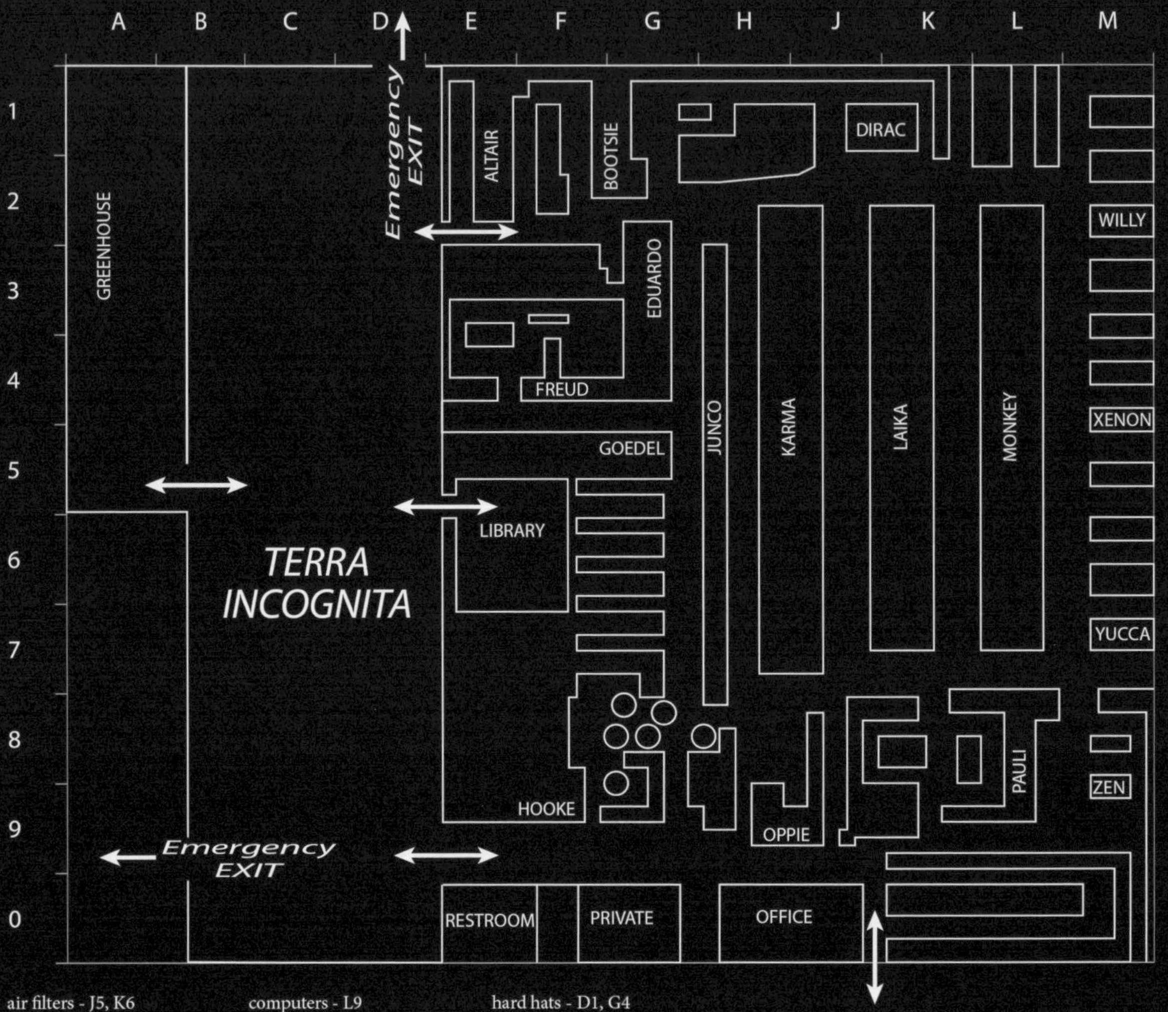

air filters - J5, K6
ammeters - G7
asbestos insulators - G3
audiotape recorders - G6
ballasts - J2
blowers - K3, K6
books - E6
bottles (plastic) - A2
cables - G1, J2
calculators - G9, H9
casters - K8, M5
centrifuge - E2
clamps, chemical - K3
clocks - D3
clothing - K9
coax cable - J2
connectors (military pinned), K4
computers - L9
couplers - M3
dewars - H1
doodads - G4
dosimeter, miitary - J4
electrical connectors - M3
electronics - G5, M1
electronics components - F4
fans - A4
fan belts - E4
film - M8
filters - F2
gas fittings - K1
gauges - K1, M2
gizmos - H7
glass - J3
graphite bricks - A1
greenhouse - A1-A5
hard hats - D1, G4
heaters - K6
hardware - G7, G8, J8
ibm cards - A1
lamps - A1, J4, J5
library - E6
light bulbs - G8
metal, scrap - D3
meters - L1, M2
motors - H5, J3-J7, L2
mylar (rolls) - D2
office supplies - L7, L9, M8, M9
optics - H9, J7, L5
philosopher's stone - E4
plotter - G3
plugs - M3
plumbing - M2
printers - L8
power cords - G8
power supplies - F6, J6
projectors - J1, J8
projector screens - K7
pulleys - M5
pumps - J4, J5
racks - E2
racks (chassis) - C7
radio equipment - G4, G5
record players - G6
records - E6
relays - J3
resistors - J3
restroom - E0
shelving - A1, D2, K1
sliders (rack, shelf) - K4
soldering irons - G7
speakers - J1
springs - F2
stereo equipment - G5
switches - M3
tape recorders - G6
thermal - M6, M7
transformers - L1
thermometers - H8
uni-strut hardware - M4
universal solvent - D9
valves - E4
valves - E4, K1
variacs - L1
voltmeters - J6
wheels - K8, M5

1984
SKI TEXAS!
PAUL HARVEY CENSORS
BUMPER STICKERS
ASSASSINATE THE CIA
DEVELOP
The Legendary

Tel. 638025, 638026
638027, 638028
Cable: IRANATOM Tehran
P.O. Box 3327 Tehran - Iran

Atomic Energy Organization of Iran
Nuclear Research Centre

Your Ref:
Our Ref:

Date: June 15, 1976

Mr. Grothus
Los Alamos Sales Co.
Box 795
Los Alamos, New Mexico 87544

Dear Mr. Grothus:

Thank you for the time you spent with me during my recent visit of Los Alamos. It was a great pleasure to see you and your warehouses of surplus equipment.

I have requested for a number of proforma invoices for the capacitors, as as soon as I receive them I will send you copies for comparison. I also visited a firm in Los Angeles, C&H Sales, 2176 E. Colorado Blvd., Pasadena, California 91107. Their prices for the capacitors is based on 6 to 9 cent per joule. You may get in touch also with them. Their telephone is (213) 6814925.

As I told you, we are willing to buy almost all your capacitors and also many other of your parts, if the price is right.

Enclosed please find $40 check I owe you for the balance of purchases I made. Please give my regards to Joe Mather, I am writing a separate letter to him.

Best regards,

M. Naraghi

M. Naraghi

MN/ar

638026
638027 638028
IRANATOM Tehran
Box 3327 Tehran · Iran

Atomic Energy Organization of Iran
Nuclear Research Centre

Your Ref:
Our Ref:

Date: Aug. 30, 1976

Los Alamos Sales Co.
P.O. Box 795
Los Alamos, New Mexico 87544

Attn: Mr. E. B. Grothus

Dear Mr. Grothus:

We would like to order some more of your surplus equipment. Would you kindly send us a proforma invoice with ocean freight charges for the following items:

1. quantity at least ten times of the items I purchased on my recent visit of your warehouse and are itemized in a few packing lists as dated May 29, 1976 (meters, connectors, capacitors, resistors, etc.)

2. Thyratron tubes all types as many as you can find with sockets, also RF power tubes.

3. High voltage power supplies 5 to 60 KV at least total quantity of 5.

4. Electrostatic voltmeters, if you have any in good working condition.

5. 4 to 20 KV storage capacitors as many as you can find.

6. High voltage relay switches, etc.

7. Miscellaneous electronic and electrical equipment, if you could get Dr. Mather's advice.

Please give us your very low prices so we may be able to buy your surplus equipment in larger volumes.

Very truly yours,

M. Naraghi

M. Naraghi

MN/ar

Don: Naraghi has been here. He seems to be kind of achisler. I don't undeßstand what kind of a racket he might have going on this stuff. Supposedly he works full time for the AEC of Iran but I know he XXX has other interests.

DO NOT TAKE ANY TUBE OUT OF THESE PLUG-IN

DRINKING WATER
DRINKING WATER
D/A CONVERTER
D/A CONVERTER
GROUNDED
Relays
Solid State
Relays
Adjustable
Time Delay
Relay
Open Armature
Relays
Advance
Relays
Clare
Western Electric
TR5A
Relay

TAKE COVER
SECOND GRADE SCHOOL CHILDREN ARE TAUGHT TO TAKE COVER IN THE EVENT OF A SNEAK ATTACK. NEW YORK CITY, 1950
PHOTOGRAPH BY SONDAK-KEYSTONE PICTURES

12/8/85

Dear Dad -

I got this postcard yesterday down at a shop in the public market, thinking it was another good anti-nuclear message you could sell at the Shalako Shop. I hope all is well. Sorry I can't make it this year at Christmas.

Love,
Tom

Mr. E. B. Grothus
The SHALAKO SHOP
P.O. Box 970
Los Alamos, NM
87544

©PHOTOWORLD
© FOTOFOLIO, BOX 661 CANAL STA., NY, NY 10013
A48

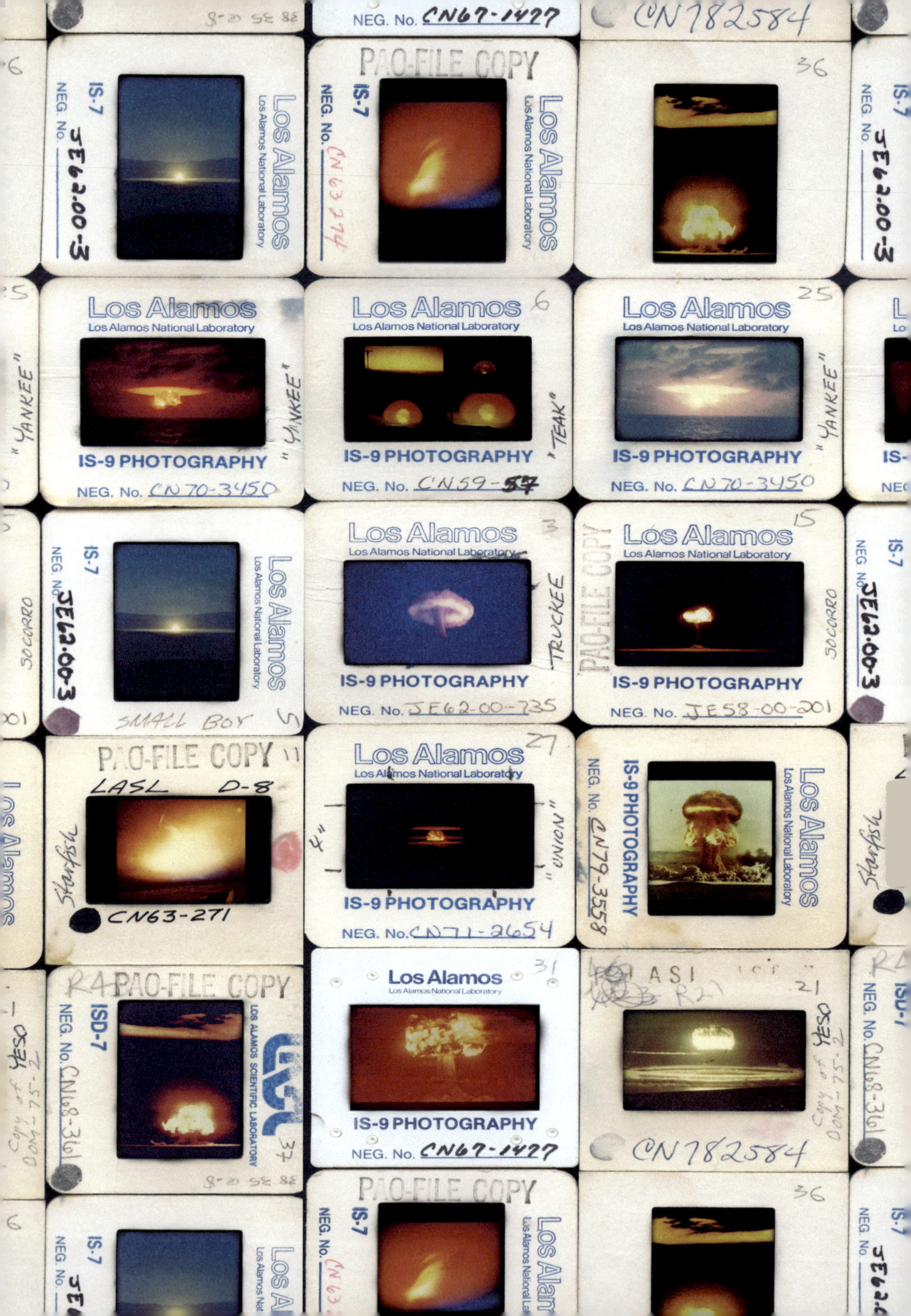
NEG. No. CN67-1427
CN782584
IS-7
NEG. No. JE62.00-3
Los Alamos
Los Alamos National Laboratory
PAO-FILE COPY
NEG. No. CN63274
36
"YANKEE"
IS-9 PHOTOGRAPHY
NEG. No. CN70-3450
"TEAK"
NEG. No. CN59-57
25
SOCORRO
SMALL BOY
TRUCKEE
NEG. No. JE62-00-735
NEG. No. JE58-00-201
LASL D-8
Starfish
CN63-271
"UNION"
NEG. No. CN71-2654
NEG. No. CN79-3558
ISD-7
NEG. No. CN68-3161
LOS ALAMOS SCIENTIFIC LABORATORY
Copy of YESO DOM-75-2
R27

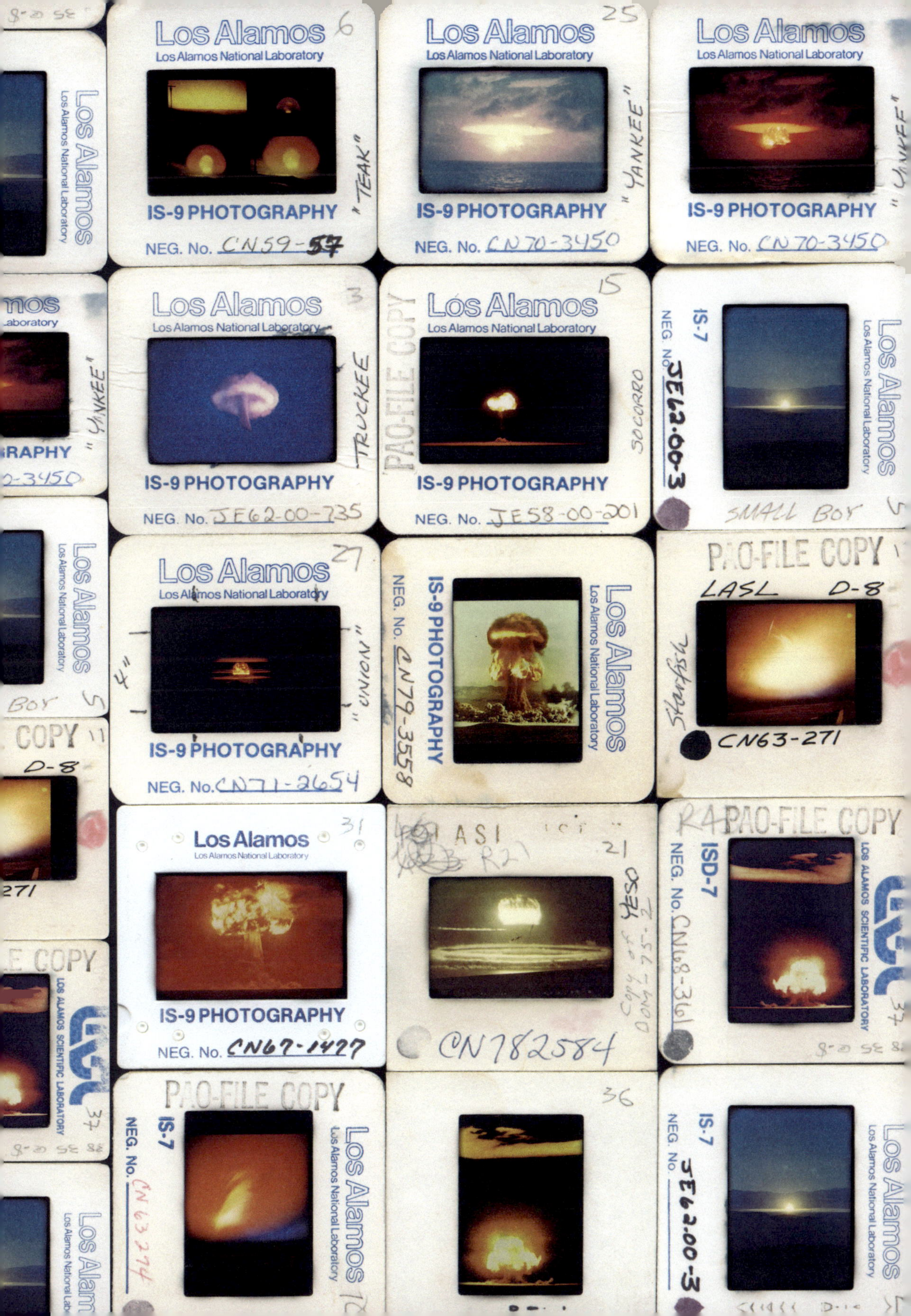
Los Alamos
Los Alamos National Laboratory
IS-9 PHOTOGRAPHY
NEG. No. CN59-57
"TEAK"
Los Alamos
Los Alamos National Laboratory
IS-9 PHOTOGRAPHY
NEG. No. CN70-3450
"YANKEE"
Los Alamos
Los Alamos National Laboratory
IS-9 PHOTOGRAPHY
NEG. No. CN70-3450
"YANKEE"
Los Alamos
Los Alamos National Laboratory
IS-9 PHOTOGRAPHY
NEG. No. JE62-00-735
TRUCKEE
PAO-FILE COPY
Los Alamos
Los Alamos National Laboratory
IS-9 PHOTOGRAPHY
NEG. No. JE58-00-201
SOCORRO
IS-7
NEG. No. JE62-00-3
Los Alamos
Los Alamos National Laboratory
SMALL BOY
Los Alamos
Los Alamos National Laboratory
IS-9 PHOTOGRAPHY
NEG. No. CN71-2654
"UNION"
NEG. No. CN79-3558
IS-9 PHOTOGRAPHY
Los Alamos
Los Alamos National Laboratory
PAO-FILE COPY
LASL D-8
Starfish
CN63-271
Los Alamos
Los Alamos National Laboratory
IS-9 PHOTOGRAPHY
NEG. No. CN67-1427
YESO
CN782584
PAO-FILE COPY
ISD-7
NEG. No. CN68-361
LOS ALAMOS SCIENTIFIC LABORATORY
PAO-FILE COPY
IS-7
NEG. No. CN63274
Los Alamos
Los Alamos National Laboratory
IS-7
NEG. No. JE62-00-3
Los Alamos
Los Alamos National Laboratory

1 Kiloton bomb
new!
Cruise Missle Warhead
B61 Warhead
high power!
Radio Frequency
NONE of the ABOVE
NPT
-CTB
O

C.T.B.T.
PRESIDENT
KENNEDY
LEFT IT...
FOR YOU
R TESTING
OT COVERED
NATIONAL
ARE PLAN
NUCLEAR
TESTING
IS A LAB
WELFARE
SCAM
UCLEAR TEST:
OW MANY
BOMBS DOES
IT TAKE TO
START AN
ARMS RACE

Mrs, Suzuko, Terada.
2230, Kuraji, Katano, City, 576.
Osaka. Japan,

VIA AIR MAIL

Mr, Edward. B. Crothus,
Box 795.
Los Alamos, New Mexico
87544. U, S, A.

Dear Edward B. Crothus

It was the most memoriable day I met you in Los Alamos on May, 6th. My daughter, who was seventeen years old, and a brother-in-law were killed in the atomic-bomb, so I hated Americans and cursed the war. As I was graduated from a mission school, I had many American teachers and some friends. I little dreamed of going to America as a member of a movement for Anti-nuclear weapon. I believed it was God's directions. I cannot help thinking that it was the Providence of God that my appeal against A-bomb was heard by you, though you made the atomic-bomb.

When you embraced me, I could hear your heartbeat and I felt your excuse and my forgiveness, then I forgot sorrow for my daughter's death with tears.

The necklace for an honorary citizen of New Mexico which you presented me becomes my treasur, and I would like to wear it forever and ever,and to give it to my daughter when I'll meet her in heaven. When I spoke about my daughter, I used to shed tears, but I decided to live for the rest of my life as a member of a movement for Anti-nuclear weapon and Anti-war, or as a witness to the war in order that no one should become a victim of war again.

I hope that we go on with a movement for peace together.

Yours sincerely,

~~Suzuko Terada~~

Suzuko, Terada

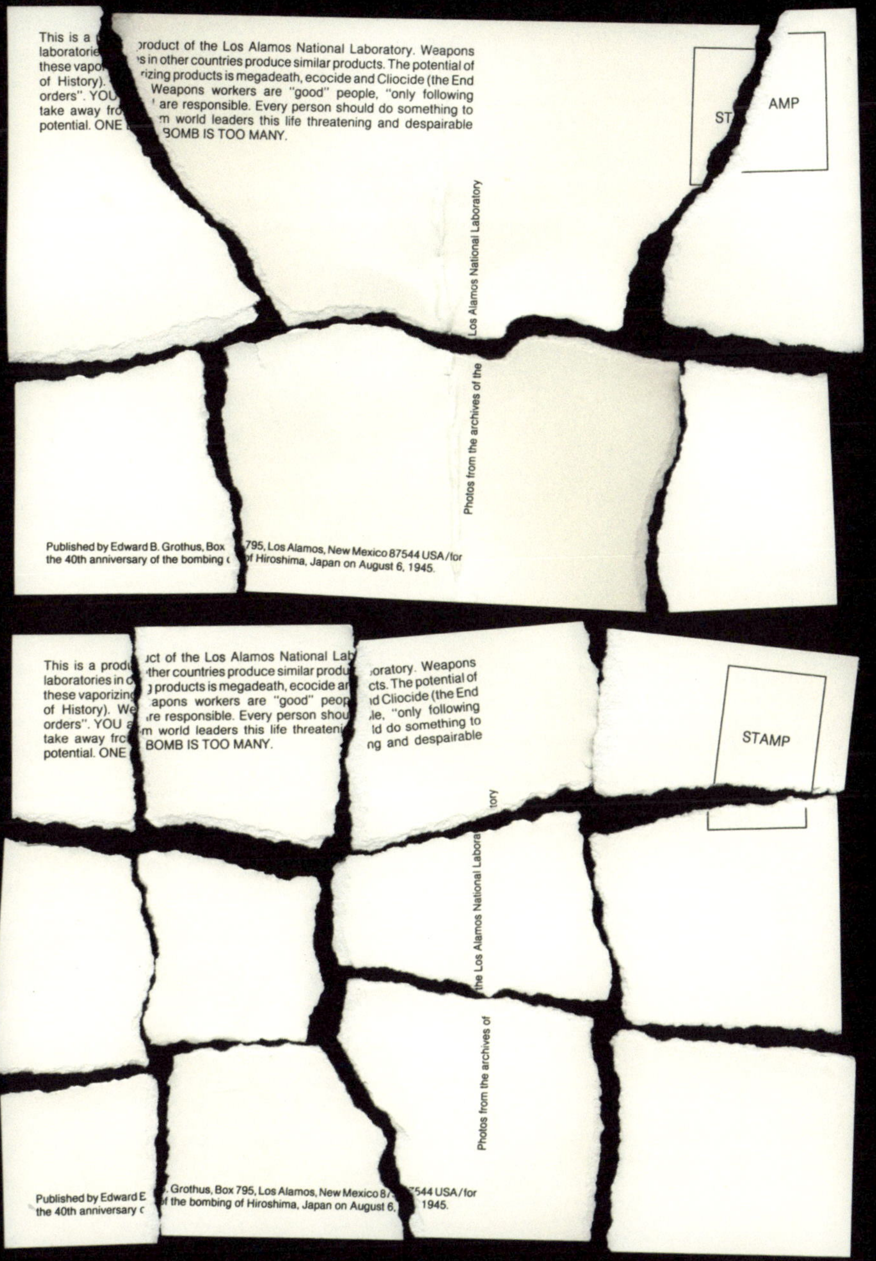
This is a product of the Los Alamos National Laboratory. Weapons
laboratorie s in other countries produce similar products. The potential of
these vapor rizing products is megadeath, ecocide and Cliocide (the End
of History). Weapons workers are "good" people, "only following
orders". YOU are responsible. Every person should do something to
take away from world leaders this life threatening and despairable
potential. ONE BOMB IS TOO MANY.
ST AMP
Photos from the archives of the Los Alamos National Laboratory
Published by Edward B. Grothus, Box 795, Los Alamos, New Mexico 87544 USA/for
the 40th anniversary of the bombing of Hiroshima, Japan on August 6, 1945.
This is a produ ct of the Los Alamos National Lab oratory. Weapons
laboratories in other countries produce similar produ cts. The potential of
these vaporizing products is megadeath, ecocide and Cliocide (the End
of History). Weapons workers are "good" people, "only following
orders". YOU are responsible. Every person should do something to
take away from world leaders this life threatening and despairable
potential. ONE BOMB IS TOO MANY.
STAMP
Photos from the archives of the Los Alamos National Laboratory
Published by Edward B. Grothus, Box 795, Los Alamos, New Mexico 87544 USA/for
the 40th anniversary of the bombing of Hiroshima, Japan on August 6, 1945.

Edward Bernard Grothus
Post Office Box 795
Los Alamos, New Mexico 87544
(505) 662-7438

December 20, 1996

Vice President Albert Gore
The White House
Washington, D.C. 20500

Dear Al,

I have been selected to help the Department of Energy dispose of it's stock of plutonium. I am the sole source for production and distribution of the material in the can enclosed herewith. Should you open and consume the material in this can you will indeed know God. It is a heavenly ambrosia and you will walk with a halo henceforth.

Should you wish to procure more of this material so that your family and friends can also walk with a halo, you may obtain this material at $5.00 per can from the address below.

The Black Hole
4015 Arkansas
Los Alamos, New Mexico 87544

Greeting of the Season,

Edo

Don Eduardo de Los Alamos
208 Andanada
Los Alamos, New Mexico 87544

Enclosed herewith: One can Organic Plutonium, Business card.

OPI
OMEGA PEACE INSTITUTE
FIRST CHURCH OF HIGH TECHNOLOGY
BLACK HOLE SYNOD
Critical Mass Every Sunday
with
Bomb Unworship Service
DON EDUARDO de LOS ALAMOS, PASTOR

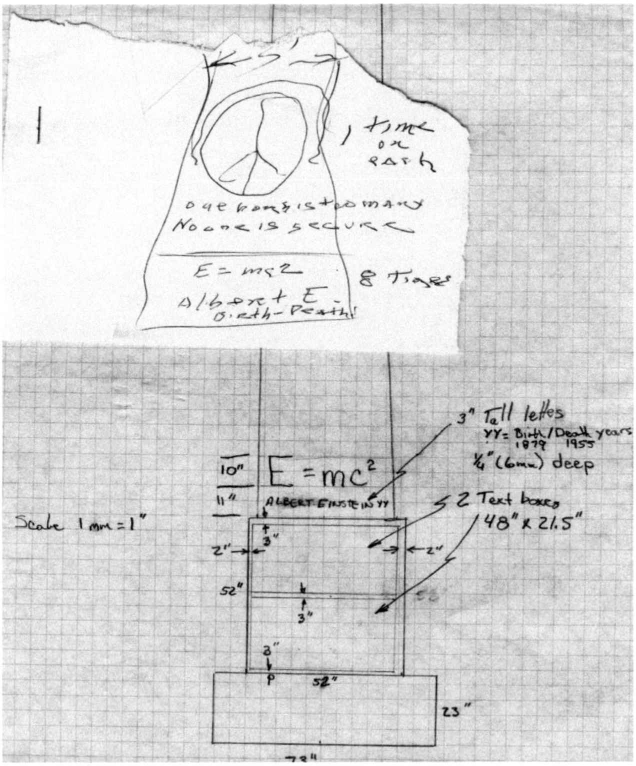
time on earth
E = mc2
3" Tall letters
YY= Birth/Death years 1879 1955
¼" (6mm) deep
10"
E = mc²
11"
ALBERT EINSTEIN YY
2 Text boxes 48" x 21.5"
Scale 1mm = 1"
52"
23"

CONFIDENTIAL

RESTRICTED DATA

This document contains restricted data as defined in the Atomic Energy Act of 1954. Its transmittal or the disclosure of its contents in any manner to an unauthorized person is prohibited.

CONFIDENTIAL

DEPARTMENT OF STATE

Washington, D.C. 20520

May 13, 1971

Dear Fellow American:

President Nixon has asked me to reply to your comments about our policy in Southeast Asia. We would like to send you a personal reply but hope you will understand that the volume of our mail sometimes forces us to use form letters.

We understand your concern about the war. One of the President's first acts after assuming office was to review US policy in Viet-Nam. He significantly changed that policy to put us on the road to peace. He has made it clear that he does not anticipate maintaining US forces indefinitely in Viet-Nam. However, the President has refused to fix a date for the withdrawal of our forces, because such a move, in the absence of North Viet-Nam's agreement to mutual withdrawals, would not bring real peace in Indochina. While a unilateral withdrawal timetable might comfort some Americans, it would offer far more comfort and encouragement to the enemy. It would remove the enemy's strongest incentive to negotiate and would give enemy commanders the exact information they need to marshal attacks against our remaining forces at their most vulnerable time.

The President's policy of withdrawing US forces steadily and by orderly stages, in accordance with the progress of Vietnamization and continuous estimates of the enemy's capabilities and intentions, but without a fixed timetable, is in our best national interests. It permits us to end our involvement while giving South Viet-Nam the chance to determine its own future.

I enclose statements about our policy and hope you will give these your full consideration. Perhaps you will then conclude as we do that the only real alternative for the United States is the fulfillment of our commitment to South Viet-Nam while pursuing our efforts to achieve a just and honorable peace.

Sincerely,

William D. Blair, Jr.
Acting Assistant Secretary
for Public Affairs

Enclosures.

MANUEL LUJAN, JR.
1ST DISTRICT, NEW MEXICO

COMMITTEE:
INTERIOR AND INSULAR AFFAIRS

JOINT COMMITTEE
ATOMIC ENERGY

WASHINGTON OFFICE:
1323 LONGWORTH HOUSE OFFICE BUILDING
AREA CODE 202: 225-6316

Congress of the United States
House of Representatives
Washington, D.C. 20515

DISTRICT OFFICES:
3006 FEDERAL BUILDING AND POST OFFICE
SANTA FE, NEW MEXICO 87501
AREA CODE 505: 988-6521

10001 FEDERAL BUILDING
ALBUQUERQUE, NEW MEXICO 87103
AREA CODE 505: 766-2538

DOUGLAS AND GRAND AVE.
P.O. 1128
LAS VEGAS, NEW MEXICO 87701
AREA CODE 505: 425-7838

September 24, 1974

Mr. Edward B. Grothus
208 Andanada
Los Alamos, New Mexico

Dear Eduardo,

Thank you for your letter and the clippings concerning the building of atomic weapons, and your views concerning President Ford's pardon of President Nixon.

The pardon is extremely controversial, and there appears to be no solution to this dilemma. Very few agree with him, but this is an Executive decision granted to the President by the Constitution.

I am very sorry we disagree over the building of atomic weapons. Many of the programs are wasteful and expensive, but I believe that the first duty of the federal government is defense. The budget for the Department of Defense, however, does not have to be treated as if it were a sacred cow, and while I believe in defense, I also contend that there are parts of this budget which could be cut.

I appreciate your taking the time to share your views, and if I can be of assistance in the future, don't hesitate to call on me.

Sincerely,

Manuel

Manuel Lujan, Jr.

ML/pap

UNITED STATES
ENERGY RESEARCH AND DEVELOPMENT ADMINISTRATION
WASHINGTON, D.C. 20545

September 2, 1977

Mr. Edward B. Grothus
Shalako Shop
Community Centre
Box 970
Los Alamos, NM 87544

Dear Mr. Grothus:

On behalf of President Carter, I am responding to your recent communication concerning the proposed LANCE nuclear warhead.

The proposed modification of the warhead for the LANCE missile system has been described as an enhanced radiation device. However, since the weapon produces much less blast and heat than a typical nuclear weapon with similar radiation characteristics, it actually reduces unwanted collateral damage to areas surrounding the target by as much as 10 to 20 times compared with the older type warheads to be replaced. The modified warhead results in fewer casualties--both friendly troops and civilians--and causes less property damage in the area surrounding the military target.

The President's stated goal is to reduce and eventually eliminate all nuclear weapons. The world situation does not permit immediate realization of that objective. A decision to produce the modified LANCE warhead would only be made to further that goal in that the same military effectiveness can be achieved with fewer weapons of smaller yields and a more modern, safe design.

The President has stated support for continuation of this program until he has been provided more information upon which to base a detailed review of the requirement to produce this modified LANCE warhead. That information has been provided and the review is underway. The concerns which you have expressed are among those which he will consider in making the final decision for production and deployment of the weapon system.

The President appreciates and respects your views, and you can be assured that they will be considered in his decision-making.

Sincerely,

J K Bratton

J. K. Bratton
Major General, USA
Director of Military Application

OFFICE OF THE ASSISTANT SECRETARY OF DEFENSE

WASHINGTON, D.C. 20301

PUBLIC AFFAIRS

2 February 1982

Mr. Edward B. Grothus
Box 795
Los Alamos, New Mexico 87544

Dear Mr. Grothus:

Thank you for your correspondence to President Reagan concerning our response to the Libyan attack on U.S. aircraft.

The United States was conducting military exercises within international waters (three miles is the recognized limit), approximately 50 miles off the coast of Libya. Thirty-nine times before the attack we had intercepted Libyan aircraft and escorted them away from the maneuvers. The fortieth time, two Libyan SU-22 aircraft made an unprovoked attack on two Navy F-14 jets. The U.S. pilots' response was a defensive action; it was either shoot or be shot down.

Our maneuvers in the Mediterranean Sea were made known to all countries in the area; hence, Libya knew that no attack was being planned and had no cause to shoot at U.S. aircraft. What we did was only a defensive measure, intended to meet our international obligation with respect to American rights abroad and in consonance with the provisions of international law.

We hope our explanation has allayed some of your concern.

Sincerely,

Philip A. Farris
Staff Assistant

ASSISTANT TO THE SECRETARY OF DEFENSE
WASHINGTON, DC 20301-3050

(ATOMIC ENERGY)

9 AUG 1989

Mr. Edward B. Grothus
Post Office Box 795
Los Alamos, New Mexico 87544

Dear Mr. Grothus:

Thank you for your letter to President Bush concerning the renaming of the town of Los Alamos, New Mexico.

The President is aware of the prime mission of the Los Alamos National Laboratory and that of its sister institutions in the Department of Energy nuclear weapons research, development, and production complex. He shares the belief of most Americans that the dedicated scientists at Los Alamos, and elsewhere, serve the nation honorably and well in their collective professional effort to ensure that the very catastrophe you fear will never materialize. Nuclear deterrence has contributed to preserving world peace for over 40 years. The fundamental credibility of this Nation's peace through strength policy is due in no small measure to the technical competence of the scientists and engineers at Los Alamos.

I share your belief that science must "embrace humanity." However, as a nation we must deter war even as we address the problems afflicting the "homeless, helpless, and hopeless." In the meantime, this Administration will continue to negotiate for verifiable arms control treaties to enhance world stability, ultimately reducing the need for all weapons.

Sincerely,

James H. Reid
Colonel, USAF
Executive Assistant

Lawrence Berkeley Laboratory
1 Cyclotron Road Berkeley, California 94720
(415) 486-4000 • FTS 451-4000

Building 70A, Room 3307
Telephone: (415) 486–5661
Telefax: (415) 486–4515

April 30, 1990

Mr. Edward Grothus, Manager
Los Alamos Sales Company
Post Office Box 795
Los Alamos, New Mexico 87544

Dear Ed:

This is in reply to your letter of April 16, 1990, and its enclosures which I found to be very informative.

I remember very well my tour of the first private housing development on Barranca Mesa soon after I assumed my duties as Chairman of the Atomic Energy Commission.

Cordially yours,

Glenn Seaborg

Glenn T. Seaborg

GTS/jl

Edward Bernard Grothus
Post Office Box 795
Los Alamos, New Mexico 87544
(505) 662-7438

August 9, 1994

Vice President Albert Gore
The White House
Washington, D.C.   2000x

Dear Vice President Gore:

In view of the upcoming conference on population being held in Cairo, Egypt I send an enclosure which may be important. If you think this information might be useful you may use it in any way.

Please keep up your good works and the works of President and Mrs. Clinton.

Kindest personal wishes,

Eduardo de Los Alamos

Enclosures: Sexual Reformation (The Grothus Creed), Copy of the PLAYBOY article of June '93, Press release.

P.S. The Vatican has not responded in any way to my mailings to the Pope.

THE VICE PRESIDENT
WASHINGTON

August 26, 1994

Mr. Edward Bernard Grothus
P.O. Box 795
Los Alamos, New Mexico 87544

Dear Mr. Grothus:

Thank you for writing to me with your thoughts concerning the upcoming International Conference on Population and Development held this September in Cairo, Egypt. I appreciate hearing from you.

I consider the dramatic growth in the world's population to be the greatest challenge currently facing the environment. As you may know, the population of the world has increased from approximately 2.5 billion people to more than 5.5 billion in less than 50 years. In effect, the population of the world has more than doubled in one generation, and it continues to steadily rise.

The effects of this rapid increase are felt around the globe. Starvation, deforestation, and lack of clean water supplies are just some of the problems facing both developed and developing countries. Certainly, finding solutions to help stabilize the world's population is one of the most crucial challenges facing all nations and one that must be addressed with innovation and determination.

I will be leading the United States delegation to the conference in Cairo in an effort to advance the goal of population growth stabilization, which is so important to the concept of sustainable development. During the conference, officials representing nations from around the globe will come together to search for international solutions to the problems associated with unchecked population growth.

The United States can play a pivotal role in changing the world's approach toward population growth. For the first time, the international community explicitly will recognize the importance of promoting education, the empowerment of women, decisions affecting family size, and ultimately decisions about the shape and nature of society itself. The conference also will recommend improvements in overall child survival rates and increased availability of family planning services.

PRINTED ON RECYCLED PAPER

Certain religious beliefs often limit the use and availability of family planning services. However, the vast majority of people in the world--of all religions--can sense the urgency and the importance of stabilizing population growth. I hope the International Conference on Population and Development will help people everywhere recognize the need to limit the size of their families.

Again, thank you for your interest in this important issue.

Sincerely,

Al Gore

AG/jec

Edward Bernard Grothus
Post Office Box 795
Los Alamos, New Mexico 87544
(505) 662-7438

December 10, 1999

President Bill Clinton
The White House
Washington, D. C. 200xx

COPY

Dear President Clinton:

The reason so many people in the world hate us is because of deterrence. Deterrence is world control by fear. Is this the way an intelligent person controls his family? You have done so much good. What would anyone gain by destroying any city anywhere in the world? This is what my country threatens to do with the Los Alamos bombs. Stockpile Stewardship is to guarantee that some future madman of our choosing will have the option to use a real, live, workable bomb at some future time. Isn't there some common sense action that you can do to stop the nuclear madness before you leave office? Please consider some action to show tolerance to those who have so little. World control by deterrence, FEAR, begets terrorists. We don't need that kind of response. We must get out of the nuclear cycle. The whole nuclear business, with unwanted nuclear wastes that no one wants is a very bad deal. PLEASE, PLEASE, THINK ABOUT WHAT I HAVE WRITTEN. You, more than anyone can make a real difference. I more than the director of the Los Alamos National Laboratory should have your ear for fifteen minutes.

Sincerely,

Ed Grothus Ed

P.S. I have never called for closure of the Lab. There are good people working here, they are just doing the wrong things. There is a fantastic infrastructure that could work for good instead of potential ill. With the Great Seal for the University of California-Los Alamos I attempt to shame the Lab into changing its mission. LISTEN TO ME.

Ed

5/7/00

Mr. Edward B. Grothus

Los Alamos NM

Dear Mr. Grothus:

For some reason, your letter of May 23 came to my attention only yesterday.

Yes, my opinion about nuclear energy has not changed. I still consider it as highly beneficent. Its safety has greatly improved in the last 20 years (it was already good then). Chernobyl was a combination of doing a stupid experiment, having incompetent people in charge, Soviet disregard for instructions (because there were too many of them), and an unstable reactor which would never have passed safety rules in Western countries.

Oil will run out, even in Arabia. There will be wars over oil, unless we have a steady, reliable source of energy, nuclear. "Renewable energy" cannot fill the demand.

Yours sincerely

Hans A. Bethe

ELMIRA NY 149
PM
08 SEP
2000

Mr. H. A. Bethe
324 Savage Farm Dr.
Ithaca, NY 14850-6503

Mr. Edward B. Grothus

THE WHITE HOUSE

WASHINGTON

October 5, 2001

Mr. Edward B. Grothus
Post Office Box 795
Los Alamos, New Mexico 87544-0795

Dear Mr. Grothus:

Thank you for your letter and your interest on issues affecting our national defense.

I believe that a strong, capable, and modern military is essential to defend our nation, advance U.S. interests, and extend our peace. Today, America's military is faced with a moment of opportunity to transform itself and thus ensure peace for generations. Seizing this opportunity will require more spending, but equally important, spending more wisely.

I have asked the Secretary of Defense to conduct a comprehensive military review to analyze the state of our military strategy, the structure of our forces, and the priorities of our budget. I want to renew the bond of trust between the President and the military by increasing pay, improving housing, and insisting that deployments have well-defined objectives.

I want to help defend against terrorist attacks at home by beefing up early intelligence, strengthening relationships with other allies, deploying anti-ballistic missile defenses as soon as possible, and sending the strong signal worldwide that assaults on our Nation's freedom will not be tolerated.

And I want to begin to create the military of the future by developing a new generation of weapons and systems that are lethal, agile, and easier to deploy. Threats to America's security must be met with swift and sure responses.

I am proud of the men and women who wear the uniform. They are on the front lines of freedom and they deserve our thanks and our commitment to their quality of life, their families, and their futures.

Thank you for writing me about this important issue. I appreciate hearing from you about your concerns.

Sincerely,

George W. Bush

Subj: Official Henry D. Smyth Report, Atomic Energy for Military Purposes
Date: Sunday, July 18, 2004 10:56:21 AM
To: edgrothus@aol.com

KNOWN - need AMs on marked ones

This book has the signatures of forty-seven of the pioneers of 1945 atomic bomb making as follows:

R. F. Bacher
Darol Froman
James T. Serdune (SERDUKE)
O R Frisch
P. Morrison
R E Schreiber

Louis Slotin

Edwin Lowry
Jack Smith
~~Julina~~ (Julius) Ashkin
Richard Ehrlich
Charles P. Baker
L H Hempleman
Mary Argo
Harold Argo
Geoffrey F. Chew
M G Holloway
Alvin C, Graves
J G Fox
Victor Weisskopf

H. A. Bethe
E J Kanopinski
N. Metropolis
Jack Calkin
(Anthony) Tony Turkovich
H S Allen
Jim Bushaman
Dave Hawkins
L. G. Parott (PARRATT)
Charles L. Critchfield
Earl A. Long
Paul C. Aebersold
Eldred Nelson (Eldred)
Robert D. Riehtmeyer
Hans Staub
LaRay B. Thompson

Harold Hirsch
James Tuck
Stanley Frankel
Leo Lavatelli
Edward Teller

J R Oppenheimer
S. F. Ulam
W Blanton Krohn
G. A. Linenberger
V. Alexander Nedgel
R. P. Feynman

47
17

Anthony Turkevich

Subj: THE BOOK
Date: Saturday, June 11, 2005 6:42:25 AM
From: EdGrothus

Dear Tom and Mike,

Thank your both for your further commentaries. I will redo the cover letter again today with your further comments in mind.

The surgeon has assured me that I will recover completely. My bowel system has almost returned to normal. I have lost another pound or two. I have had a slight hint of indigestion. I sleep better. I am lethargic. I try to remain active during the day. My hands tremble. Margaret says apples and melons are gas formers. The cool melon tastes good. I try to drink more water. I shower and shave.

Margaret drove me to the bank, the PO and by the Black Hole yesterday. Greg Mello was here yesterday with three others. One guy was the National Coordinator for Pax Christy. Greg Mello is the director of the Los Alamos Study Group which is the major watchdog group here. I had a good talk with the four. I keep up with my computer corespondents. I read the papers but not much else.

Peace and love,

Edward B. Grothus, 208 Andanada, Los Alamos, New Mexico 87544 USA
Phones: (505) 662-7438 home and (505) 662-5053 work

Subj: **THE BOOK**
Date: Wednesday, April 5, 2006 8:53:03 AM
From: EdGrothus
To: [redacted]@sothebys.com

Dear Elizabeth,

You have in your possession a book with the signatures of the men who created the most significant happening in the history of the world. You, your auctioneer and the person at Sotheby's who might write a press release to advertise the sale of the book should be completely familiar with the life and times of these men. The legacy of these men haunts the world. Will humanity survive the Nuclear Age?

Oppenheimer headed the scientific community here in Los Alamos, Project Y. Teller and Stan Ulam were cofathers of the H-Bomb but Teller did not care to share the idea with Stan. Hans Bethe was the first head of T-Division, the Theoretical Division. Otto Frisch was with Lise Meitner in Sweden when they first realized that Hahn and Strassman, in Berlin, had split an atom. Feynman was a brilliant wild card, physicist and he and Bethe were both honored with Nobel prizes. Weisskopf, yet another foreigner, collaborated with them all. Louis Slotin died tickling the dragon's tail shortly after Trinity, Hiroshima and Nagasaki. Tuck was part of the English mission in '44-'45. He and Metropolis and a few others lived and worked here until they died.

I wish you, your auctioneer and your press person could visit Los Alamos and spend a few hours with me here. Your auctioneer could well drive the hammer price to a level far beyond the early estimates. He needs to be very familiar with the history he is selling. Such an accomplishment would be feather in Sotheby's cap to be exploited for years.

I don't know your auction room setup but think the room should have opaque projectors which show the signatures. Large posters of these two pages would also be effective.

Your press release should go to all of the major newspapers in the world. London, Paris, Berlin, Rome, Tokyo, Beijing, Moscow and Copenhagen to name just a few.

Please forgive me for wanting to be a participant in this offering.

I want to write personal letters to the richest computer people who have money they do not know what to do with.

As for myself I am the most outspoken person here. I read. I write. I participate. I have lost all of my faith in the bombs and the whole nuclear business. I write and am a real pain-in-the-ass here. Because of my singularity I am the most interviewed and the most photographed person in Los Alamos. I worked in a weapons development group from '51 to '69. I did, singularly, the measurements which made "better" bombs. During my time at GMX-4 we reduced the size of the bomb by thirty times while the yield was increased by thirty times, 15 kilotons to 475 kilotons. The physics package today is the size of a bowling ball. I am a very active 82, 83 in June, the 28th. GMX, means Gadget, iMplosion, eXplosives.

Peace and love,

Edward B. Grothus, 208 Andanada, Los Alamos, New Mexico 87544 USA
Phones: (505) 662-7438 home and (505) 662-5053 work
edgrothus@aol.com or ed@blackholesurplus.com

Sotheby's EST. 1744
1334 YORK AVENUE NEW YORK NY 10021
Tel (212) 606-7444 Fax (212) 606-7355 WWW.SOTHEBYS.COM
Tel (212) 606-7000 WWW.SOTHEBYS.COM

Remittance Advice

MR EDWARD GROTHUS
208 Andanada Street
Los Alamos NM 87544-2404

Document Date: 26-JUL-2006
Client Account Number: 51128621
Remittance Number: 2000226374
Cheque Number: 7000010527

FORM 0511

THE FACE OF THIS DOCUMENT HAS A COLORED BACKGROUND ON WHITE PAPER

Sotheby's EST. 1744
1334 York Avenue
New York, NY 10021

7000010527

LaSalle Bank NA
Chicago, IL 60661

70-2302
719

DATE: 07/26/2006 AMOUNT: 20,355.00

PAY*** TWENTY THOUSAND THREE HUNDRED FIFTY-FIVE USD***

Pay To : MR EDWARD GROTHUS

SOTHEBY'S,

by

Authorized Signatory

SECURITY FEATURES INCLUDED. DETAILS ON BACK.

⑈7000010527⑈ ⑆071923022⑆ 0079280⑈

Subj:	**Los Alamos 12/1/07**
Date:	Saturday, December 1, 2007 7:17:36 AM
From:	EdGrothus
To:	BGrothus, EdGrothus, HGrothus, MGrothus

Dear Barbara, Tom and Wendy, Mike and Heidi,

This project could not have been completed without your help. Tom finalized the translations. Heidi found Feng. Mike and Barbara were at my service with welcome art advice and telephonic assistance. The projects conception was mine but without you all this project would NEVER have been brought to fruition.

I hope to do a press release about our project. I am kind of relaxed that everything is on site and in perfect condition. I don't know what reaction there might be nationally or internationally. (There he goes again thinking BIG.) Advise me. Talk to me.

It was the Art in Public Places Board that acted locally. I do not know if there might a different local reaction. The monuments are impressive and now I can show and tell of everything. I told the Monitor about the unloading but no one showed up yesterday. Should I make anything of that?

My criteria for location includes visibility for a great distance, accessibility so people can read the translations and security. OPI fails on the first most important score.

I understand, I think, Mike, your concerns about a location done soon. Art in a private place is possible. Might they be removed from in front of OPI to another location at some future time? Obelisks have a history of being relocated.

I may have voiced the fear to all that we might blow ourselves away BEFORE the monuments get erected. Abolition is essential for humanities survival. War begets untold death and destruction. Is a transition from war to peace possible? Can the five ruling entities learn from the experience of the current stupid and unrewarding war? The entities are:1) The civilian leadership. 2) The politicians. 3) The CEO's of the military/industrial complex. 4) The obedient people in the Pentagon and LANL. 5) The churches.

The problem to make change is enormous. Is the pen mightier than the sword? From my daily experiences at the Black Hole I know I am on the right track. The comments in my Guest Book reassure me. I recognize the onus (burden, responsibility, load) to implement peace. I am not deterred. Is it possible that the Monuments might facilitate a change in the way the powerful think? I am pleased with the words inscribed.

We can now do foundation plans. They will be useful even without a confirmed site location. They are not minimal.

Roger Lamoreaux, the craneman from Santa Fe, was most impressed and I casually mentioned I would need him to make the erections. We smiled.

It is raining and the wind is blowing. There are roof problems at the Black Hole. There is business to attend to. Winter is near. I am thinking. I am working. I will be writing. Dialog is essential. "Who knows what evil lurks in the minds of men. Only the shadow knows." I will write to Feng.

Subj:	**Monuments**
Date:	Sunday, December 2, 2007 2:07:32 PM
From:	EdGrothus

Dear Feng,

The two 20' containers have arrived in Los Alamos and were unloaded from the flatbed trucks on Friday. Everything has arrived in perfect condition. I want to check the inscriptions to see if they are all correct. The containers are used and not in very good condition. We accept them.

We will be working to find a place to erect the monuments. I will write a story about the monuments for national and international newspapers. I will send you a copy.

I tell you again how pleased I am with the work done in China. Yours is a great nation. Your history of the last fifty-eight years is remarkable. China is rising while the USA is falling. We have warmongers in charge. The war is killing the USA in many ways. I do what I can to prevent a nuclear war. A nuclear war could happen by accident or evil intent. We all live in danger of our lives.

We cannot put the genie back into the bottle but with abolition of nuclear bombs we could all be certain of a future. The future for our children is in doubt now.

I recommend your work to many others. I hope there will be more business for you.

How much does one of the 35' tall statues that we saw in Jinjang cost?

Peace and love,   Ed Grothus in Los Alamos, New Mexico, USA

Edward B. Grothus, 208 Andanada, Los Alamos, New Mexico 87544 USA
Phones: (505) 662-7438 home and (505) 662-5053 work
edgrothus@aol.com

****

2/8/71 11:50 AM

IRAQ AIR
AIR WAYBill 096-852/2083
A.W.B

9.6.78

Items	Dimension (in × in ×)	Cuft	Wt
9.16	20.5 × 20.5 × 20"	4.9 ✓	83#
4	20.5 × 21 × 12.5	3.1 ✓	54
47, 48, +	17.5 × 10 × 23.5	2.4 ✓	61#
19	21 × 18 × 16.5	3.6 ✓	50
45	17.5 × 20.5 × 13	2.7 ✓	94
50	20.5 × 21 × 10.5	2.6 ✓	85
~~20 — 27 × 25.5 × 14~~		~~5.6~~ ✓	95
12, 41, 89 +	13.5 × 25.5 × 30	6.0 ✓	72
57	24 × 21 × 12.5	3.6 ✓	83
11	20 × 29 × 21	7.0 ✓	95
59	24 × 28 × 20.5	8.0 ✓	110
42	24.5 × 23.5 × 19.5	6.5 ✓	44
17	18 × 22 × 13	3.0 ✓	29
96; 7-8 ch	7.5 × 37½ × 4½	.73 ✓	65
55	13.5 × 13 × 24.5	2.5 ✓	34
7	20.5 × 17 × 15.5	3.1 ✓	31
37	16 × 19.5 × 10	1.8 ✓	19
5	17 × 15.5 × 20.5	3.1 ✓	47
(8,5)	17 × 10 × 21.75	2.1 ✓	33
38	30 × 19.5 × 8	2.7 ✓	33
2,3	17.5 × 23 × 10	2.3 ✓	32

JANIRE NÁJERA

The Black Hole

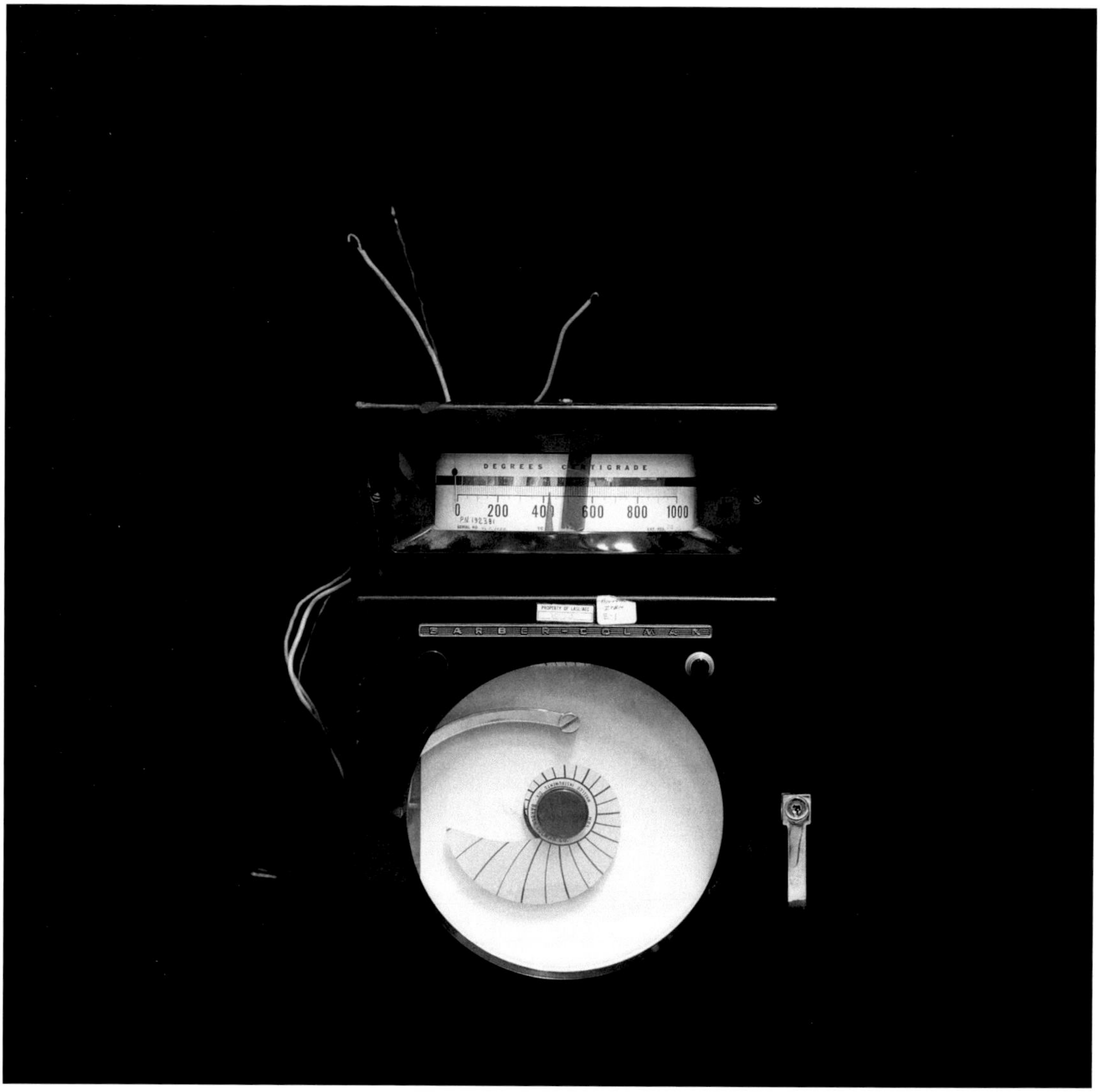
DEGREES
TIGRADE
0 200 400 600 800 1000
BARBER-COLMAN

LASL
SCOPE CAMERA
MAGAZINE
No.106

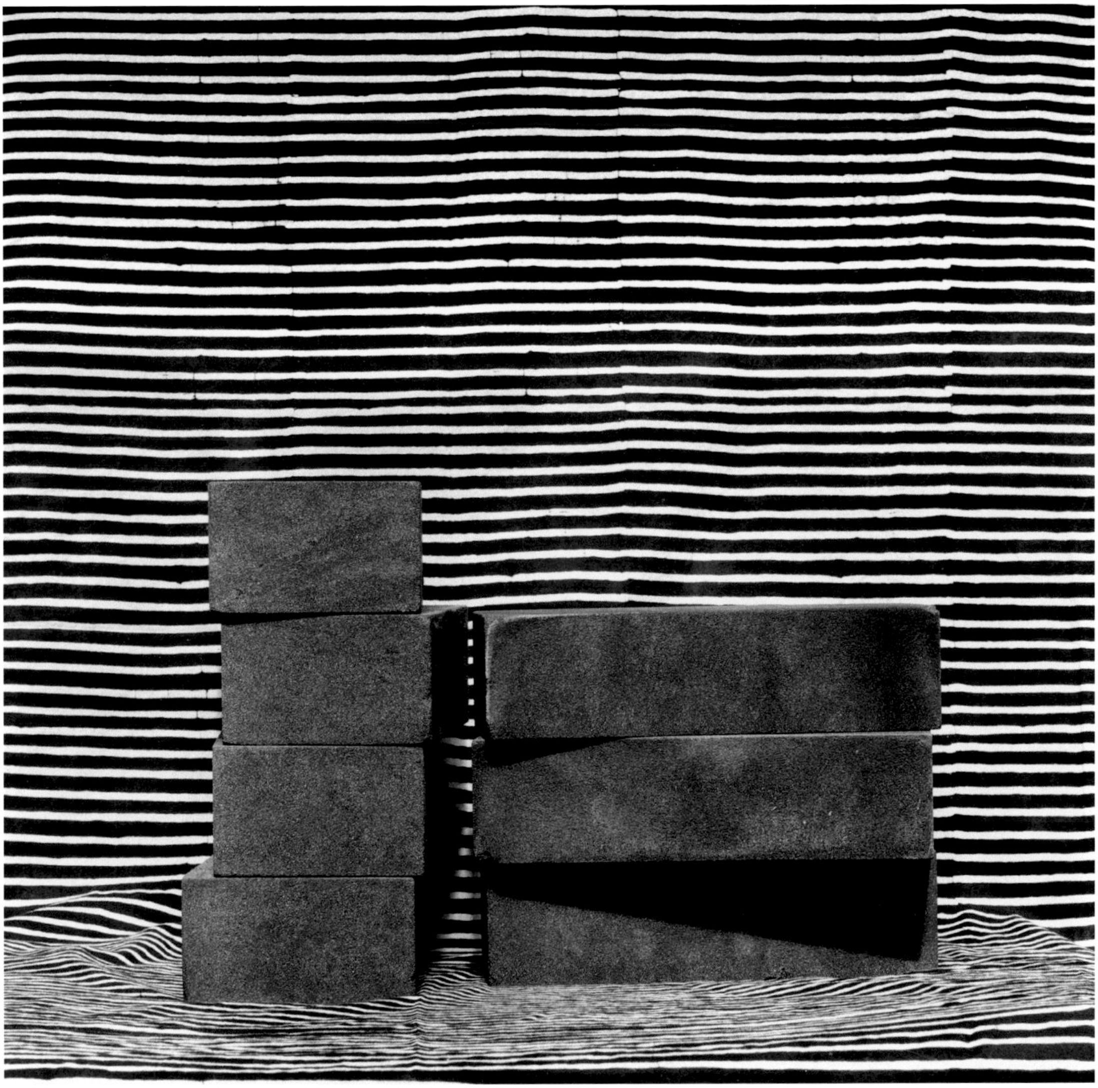

CMB 6 2/2/ 68
L A S L

The Living World
GEORGE B. JOHNSON
Nuclear Power Issues and Choices
Report of the Nuclear Energy Policy Study Group
Spurgeon M. Keeny, Jr.
CHAIRMAN
Seymour Abrahamson
Kenneth J. Arrow
Harold Brown
Albert Carnesale
Abram Chayes
Hollis B. Chenery
Paul Doty
Philip J. Farley
Richard L. Garwin
Marvin L. Goldberger
Carl Kaysen
Hans H. Landsberg
Gordon J. MacDonald
Joseph S. Nye
Wolfgang K. H. Panofsky
Howard Raiffa
George W. Rathjens
John C. Sawhill
Thomas C. Schelling
Arthur Upton
Sponsored by the Ford Foundation
Administered by The MITRE Corporation

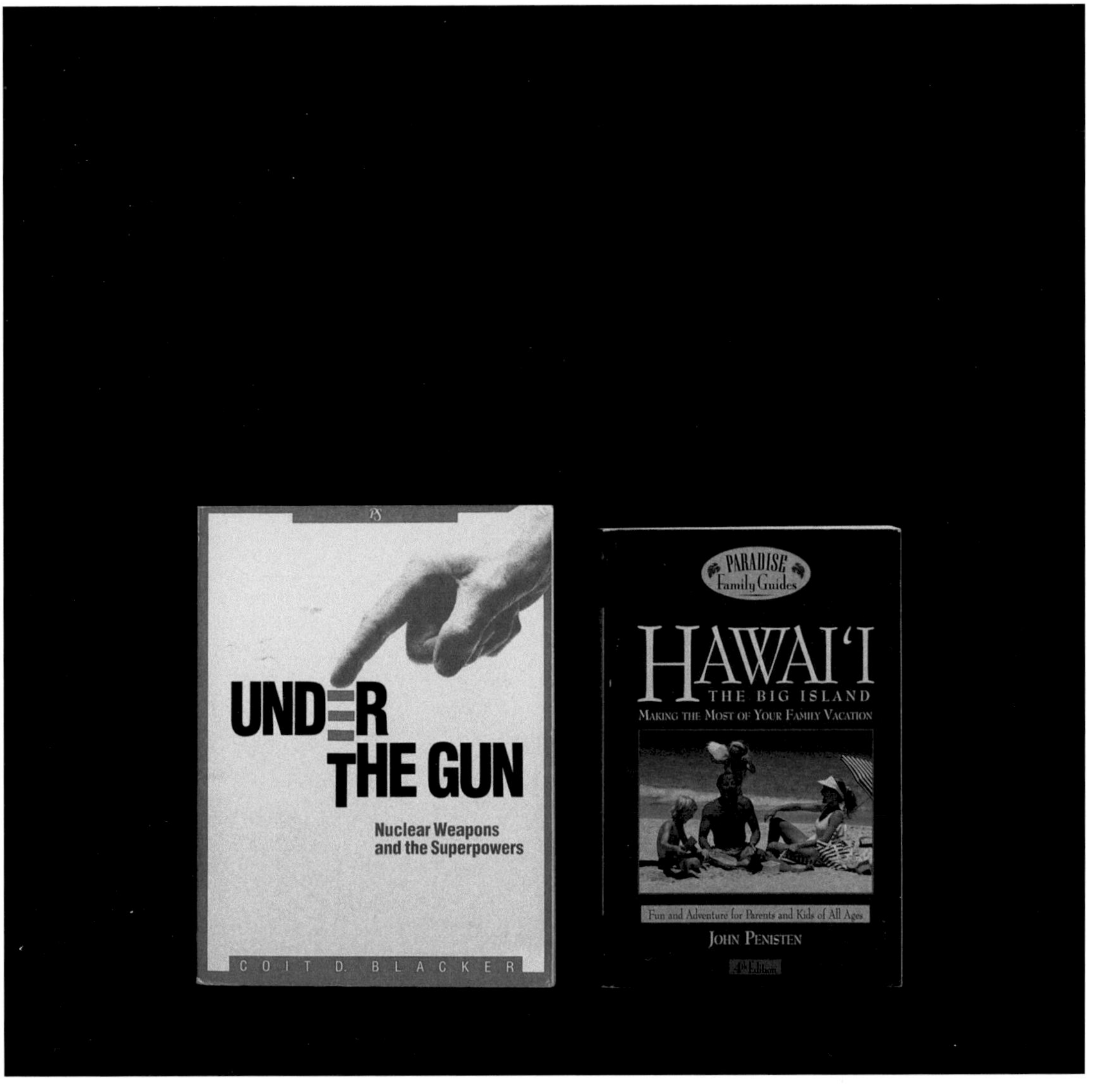
UNDER THE GUN
Nuclear Weapons and the Superpowers
C O I T D. B L A C K E R
PARADISE Family Guides
HAWAI'I
THE BIG ISLAND
MAKING THE MOST OF YOUR FAMILY VACATION
Fun and Adventure for Parents and Kids of All Ages
JOHN PENISTEN

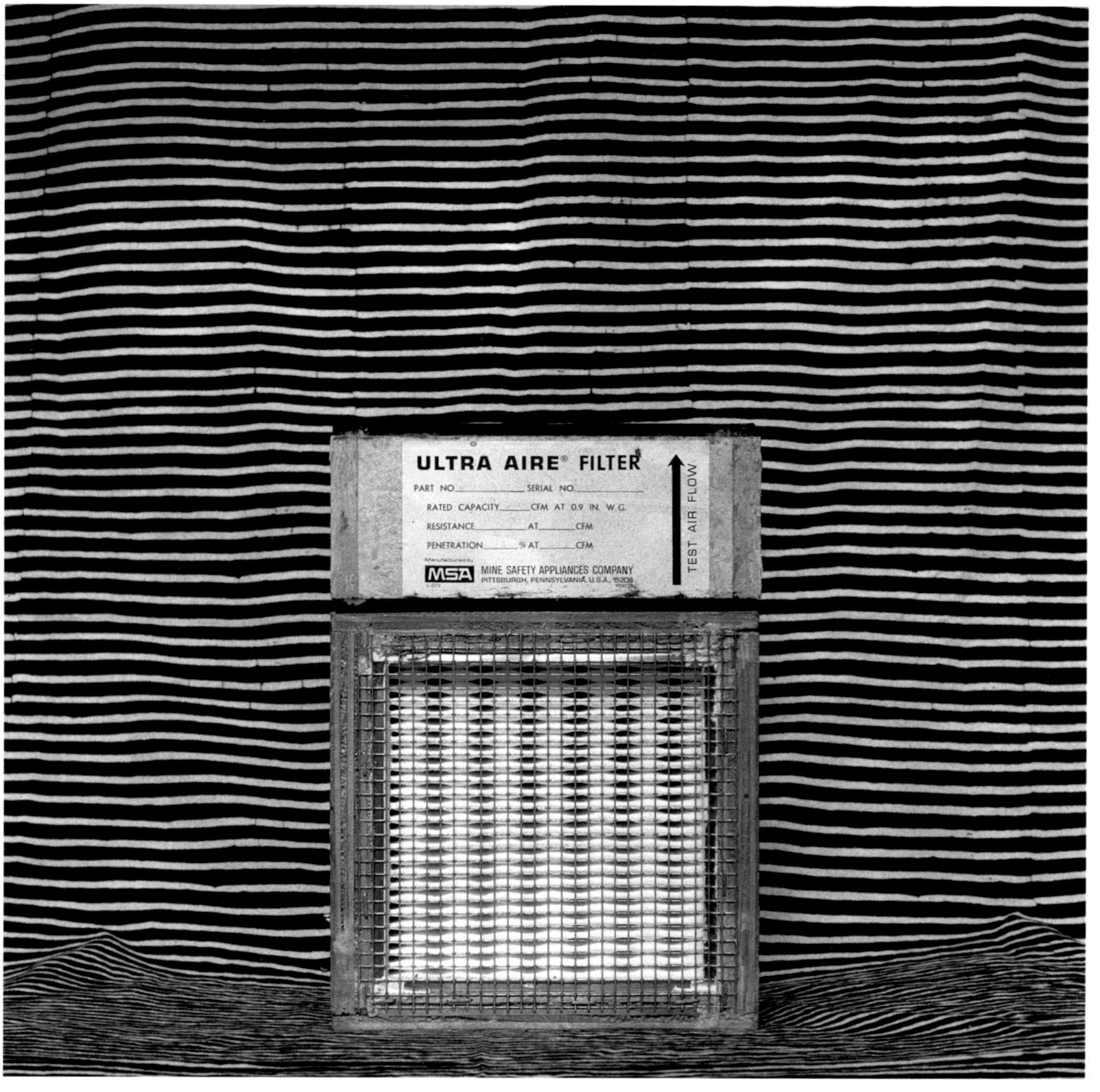
ULTRA AIRE® FILTER
PART NO. SERIAL NO.
RATED CAPACITY CFM AT 0.9 IN. W.G.
RESISTANCE AT CFM
PENETRATION % AT CFM
MSA MINE SAFETY APPLIANCES COMPANY
PITTSBURGH, PENNSYLVANIA, U.S.A., 15208
TEST AIR FLOW

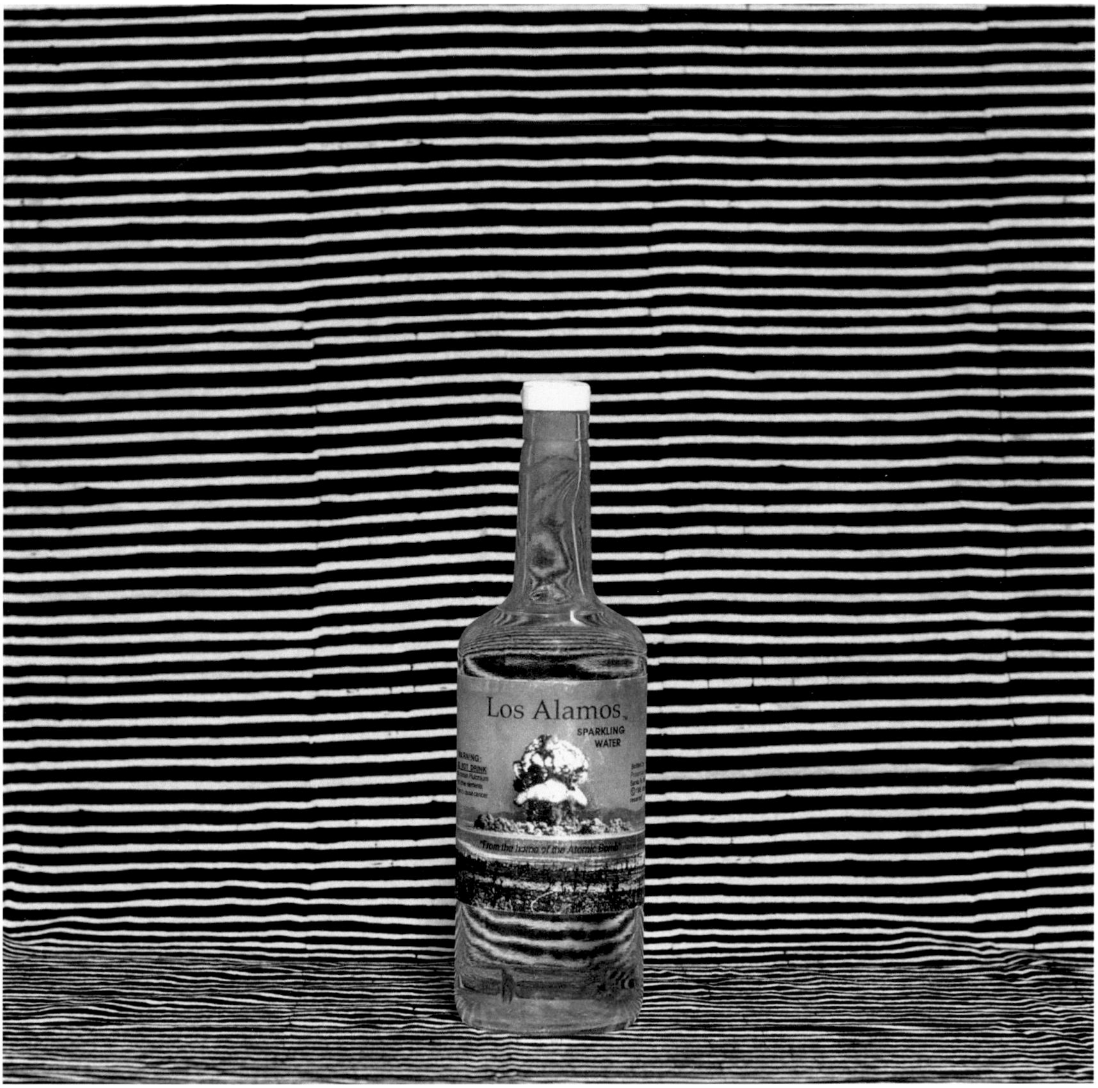
Los Alamos
SPARKLING
WATER

CAUTION
UNPROTECTED GLASS DEWAR!

Bell & Howell
FRAMER →
IMPORTANT!
BEFORE ENGAGING FILM TO TAKE UP SPROCKET (RED ARROW) DRAW FILM TIGHT TO LEFT, TO TAKE OUT ALL SLACK THEN RELEASE FILM UNTIL FIRST AVAILABLE SET OF FILM PERFORATIONS ARE IN POSITION FOR ENGAGEMENT WITH SPROCKET TEETH (CORRECT SLACK IS EQUIVALENT TO 1/2 FRAME.

CAUTION
RADIATION
HAZARD

ECI
CALIBRATION LABORATORY
INSPECTION & TEST EQUIPMENT
4400006
4400006

1
2
3
4
5
6
7
8
9
10
11
12
13
14
15
16
17
18
19
20
21
22
23
24
25
26
27
28

TEXXOR
heat cell

F&M Scientific Corporation
COOL

KW DEMAND
KILOWATT HOURS
MULTIPLY BY1000
KW DEMAND
KILOWATT HOURS
MULTIPLY BY1000

Bell & Howell

Burroughs

HYGRO-THERMOGRAPH
RECORDING
RELATIVE HUMIDITY
AIR TEMPERATURE

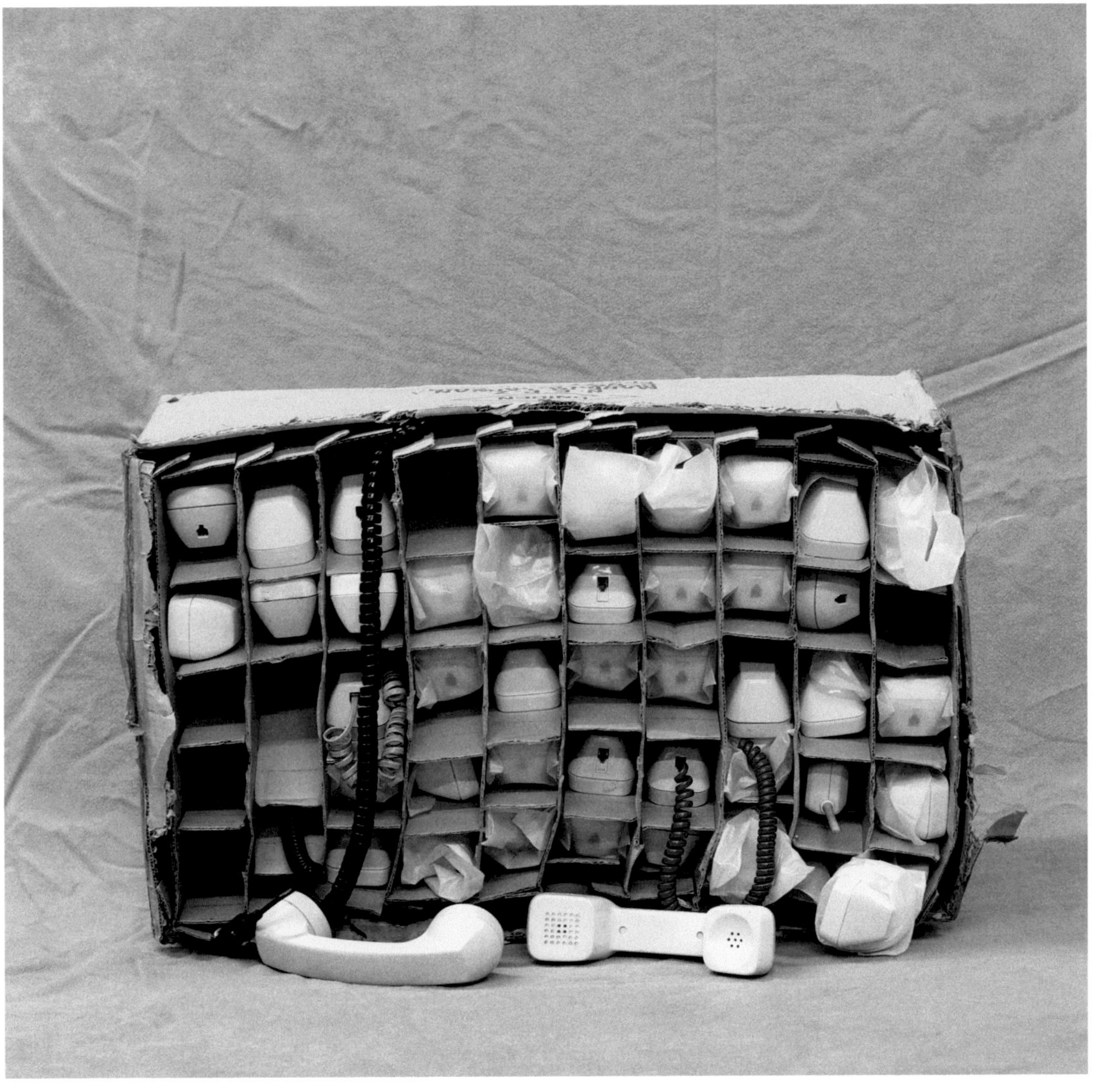

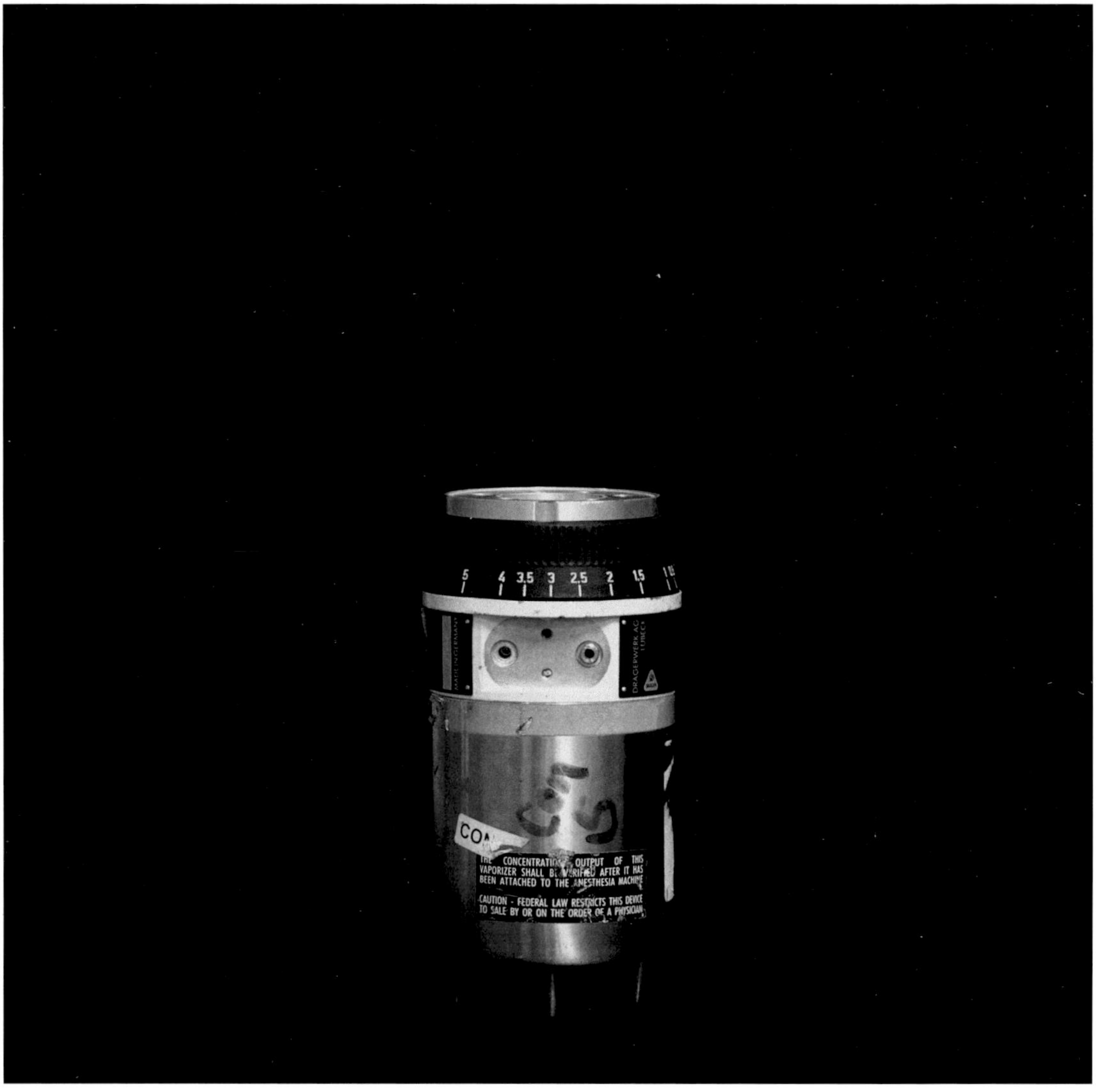
5 4 3,5 3 2,5 2 1,5
MADE IN GERMANY
DRAGERWERK AG
CO
OUTPUT OF THIS
VAPORIZER SHALL
AFTER IT HAS
BEEN ATTACHED TO THE
CAUTION - FEDERAL LAW RESTRICTS THIS DEVICE
TO SALE BY OR ON THE ORDER OF A PHYSICIAN

NITROGEN
REFRIGERATED LIQUID
UN 1977
ALWAYS KEEP CONTAINER IN UPRIGHT POSITION.
WARNING
Extremely cold liquid and gas.
Can cause rapid suffocation.
May cause severe frostbite.
NON FLAMMABLE
FIRST AID
DO NOT REMOVE THIS PRODUCT LABEL
Minnesota Valley Engineering, New Prague, Minnesota 56071, U.S.A.

A·B·E
"700"

In the final chapter of his book *Photography Today,* Mark Durden describes a provocative photographic installation, comprising roughly 350,000 individual prints and entitled *24 Hrs in Photos*:

> In November 2011, Erik Kessels filled an Amsterdam gallery with prints of images uploaded to Flickr over twenty-four hours… physically bringing into the gallery the problem and issues of mass photography, a photography that normally circulates virtually, from screen to screen.[1]

The words "problem and issues" are significant in that they reference a common criticism levelled at contemporary image-making (and image-makers): that they are superficial, self-absorbed, lacking in discernment, and worse. Kessels himself alludes to this, saying of his mountainous piles of prints that he aimed to "visualize the feeling of drowning in representations of other people's experiences."[2] It is the case that the billions of images travelling instantaneously back and forth every second of every day glut our conscious (and unconscious) minds, whether or not we are aware of the fact. As long ago as 1994, Sven Birkerts was unequivocal about the likely consequences of this phenomenon: "We are poised at the brink of what may prove to be a species mutation. We had better consider carefully what this means."[3]

This kind of negative commentary is not a recent thing: as Fred Ritchin points out, it is "in fact reminiscent of earlier critiques of photography, such as Charles Baudelaire's famous 1859 remark: 'If photography is allowed to supplement art in some of its functions, it will soon have supplanted or corrupted it altogether, thanks to the stupidity of the multitude which is its natural ally.'"[4] So, even in the mid-nineteenth century, people were grumbling about the degradation of "art" through the use of mechanical devices; and as the technology evolved and cameras became accessible to "the multitude," such essentially elitist complaints were voiced more widely. It is ironic, therefore, that Baudelaire was also one of those responsible for the reductive theory that poetry was *assembled*, rather than created by the writer: "As Baudelaire rightly said, the poet is a compositor, the poem is essentially composed and composite. The distinct elements of the text are only apparently placed end to end."[5]

Given that he held such a mechanistic view of the production of poetry, it is likely that Baudelaire would approve of the kind of assembling, reworking, and recontextualizing of photographic images practised by Kurt Schwitters, the Dadaists, and many more, throughout the twentieth and early twenty-first centuries. They operate according to a different strategy from the profligacy of Kessels's (admittedly egalitarian) work: those who adopt a more disciplined approach to the sorting, sifting, and recycling of existing photographic imagery often do so partly in response to what they regard as the surfeit of images already in the world. Fred Ritchin describes these folks thus: "There are others who have moved into a role not unlike that of an archivist, or a 'metaphotographer,' gathering work from people who, from direct experience, may know the most about a situation."[6]

While Ritchin cites the work of Susan Meiselas and Geert van Kesteren as examples, Janire Nájera's six-year project, whose outcomes include *The Black Hole* and *Atomic Ed,* can equally well be described as the work of a metaphotographer. Her methodology involves a commitment, a depth of engagement and an immersion in the subject(s) that goes far beyond the simple —or even complex— strategies of appropriation and re-presentation. Writing in 2017, the writer and curator Daniel Palmer expanded upon Ritchin's definition of the metaphotographer to include those who, like Nájera, work with found imagery: "They sort through photographs that already exist in the world, and make those images visible in a different way by providing them with a new context. They aggregate, which is to say they collect related photographs so as to display their connections."[7]

The practice of Nájera and others like her, who work with found and vernacular imagery in addition to creating their own, embodies a compassion and a generosity of spirit to which Palmer also makes oblique reference. He writes that "by publicly recontextualizing personal photographs... artists draw out unconscious narratives within collections of amateur images, reinvesting them with a different authority and pathos in what I take to be a collaborative rather than simply appropriative act."[8] There is no doubt that, throughout the first section of this book, we can sense the presence of its eponymous subject in all his quirkiness, passion, and doggedness, and Nájera's adroit assembling of elements from an array of sources succeeds in creating a nuanced and multi-layered portrait of Ed Grothus.

The second section of the book comprises Nájera's series of photographs *The Black Hole.* It constitutes a taxonomy as remarkable for its opacity as that found within the pages of the book *Evidence,* created by Larry Sultan and Mike Mandel. Their audacious project is summarized by Sandra Phillips as "a sort of funny referendum on the new technology, burgeoning close by in the incipient Silicon Valley."[9] The subjects —both human and inanimate— of this parade of black-and-white images appear preposterous and pathetic in equal measure, and sometimes both at once. The result is "an intuitive visual essay about the ridiculousness of vanity… [that] makes the pictures themselves essentially unknowable."[10] It is a book created by two early metaphotographers that tells us as much about the slippery yet powerful medium of photography as it does about its subjects.

Like Sultan and Mandel, Nájera presents the viewer with often inscrutable representations of objects whose precise function evades us, even as we are drawn to look more closely. Some pictures are easy to read, if slightly odd: a bouquet of Old Glories standing against a striped backdrop, for example; while others display what seems to be equipment for measuring, magnifying, and observing. There is a further category of images that remain impenetrable, confounding all but the specialist viewer. Given that we are taught that monochrome photography, especially, offers us so much in the way of clarity, realism, and (even today) truth, this is a cause of some frustration. It is, however, just the kind of cryptic puzzle that would have tickled Ed Grothus no end.

As was the case with the first section of *Atomic Ed,* its inimitable subject seems to stand close enough to peer over our shoulder as we turn the book's pages. One can almost hear him chuckle as we pore over its often mysterious contents. Daniel Palmer again: "The original authors and even subjects of found personal photographs continue to exist as absent presences, and therefore as unconscious collaborators."[11] I'm left with the vision of Ed Grothus wrapping up copies of the book to send to leading politicians all around the world, and therefore end with one of his favourite sign-offs: "Peace and love."

1. Mark Durden, *Photography Today: A History of Contemporary Photography* (London & New York: Phaidon, 2014), p. 448.
2. Ibid., p. 448.
3. Sven Birkerts, *The Gutenberg Elegies: The Fate of Reading in an Electronic Age*, 2nd ed. (London: Faber and Faber, 1996), p. 31.
4. Fred Ritchin, *Bending the Frame: Photojournalism, Documentary, and the Citizen* (New York: Aperture, 2013), pp. 47-48. The quotation from Baudelaire is taken from "The Salon of 1859" [first published in French in the *Revue française* (Paris), June 10-July 20, 1859], in *The Mirror of Art: Critical Studies by Charles Baudelaire*, ed. and trans. by Jonathan Mayne (London: Phaidon, 1955).
5. Pierre Macherey, *A Theory of Literary Production* (London and New York: Routledge, 2006) [first published in French in 1966].
6. Ritchin, *Bending the Frame*, p. 38.
7. Daniel Palmer, *Photography and Collaboration: From Conceptual Art to Crowdsourcing* (London and New York: Bloomsbury Academic, 2017), p. 169.
8. Ibid., p. 138.
9. Sandra Phillips, in Larry Sultan and Mike Mandel *Evidence*, 2nd ed. (New York: Distributed Art Publishers, 2004).
10. Ibid.
11. Palmer, *Photography and Collaboration*, p. 143.

Edward Bernard Grothus

A look through the archive of the legendary Atomic Ed.
All images are from Ed Grothus's archive.

— Page 3
Ed Grothus's father, Edward Theodore Grothus. Photo taken in 1964.

— 4
Ed as an infant in Iowa.

— 7
Ed's mother, Regina Hebinck G. Photo taken in 1964.

— 8 — 9
A young Ed Grothus on one of his trips to the Caribbean and South America. Rio de Janeiro, Brazil, 1947.

— 10 — 11
Pages from Ed's private diary, kept from March 24th, 1946 to March 25th, 1950.

— 12
Ed Grothus in Los Alamos, two years after his arrival in New Mexico.

— 14 — 15
Offer of employment from the Los Alamos Scientific Laboratory (LASL). 1949.

— 16
Reminder letter from the LASL. 1949.

— 17
Employment confirmation letter from the LASL. 1949.

— 19
Atomic Energy Commission identity card.

— 20 — 21
Archival image of Los Alamos, marked up by Ed Grothus. 1951.

— 22
Group portrait of employees at the twenty-year reunion of the Los Alamos Scientific Laboratory. November 19th, 1969.

— 23
Ed with fellow protesters at Los Alamos, displaying one of his favorite slogans.

— 24 — 25
Illustration from the booklet *You and the Atomic Bomb: What to Do in Case of an Atomic Attack*. New York State Civil Defense Commission.

— 26
Building the first adobe house in Los Alamos, at Barranca Mesa. 1960.

— 27 — 28
Telegrams sent to President Richard Nixon and to Mr. Henry H. Fowler. 1970/1971.

— 30
Margaret Grothus and her children inspect the timber cut for the construction of their new adobe house. 1960. Left to right: Barbara, Tom, Becky Cole, Suzi, Mike, and Margaret Grothus.

— 31
Letter from Ed's mother, Regina Hebinck Grothus, a World War I Navy veteran.

— 32 — 33
Telegrams sent to President Nixon and President Gerald Ford. 1972/1973/1974.

— 34
Family photograph of Ed, Margaret, Tom, Barbara, Susan, Mike, and Ted Grothus.

— 35
The adobe house with its finished exterior in 1961. The interior took another decade to complete.

— 36 — 37
Response sent to Ed Grothus from the White House. 1975.

— 38
Winter view over Los Alamos County.

— 40 — 41
Film scan found in Ed Grothus's archive.

— 42
Heidi, Mike, Meg, and Casey Grothus with the dog Dot at The Black Hole in Los Alamos.

— 43
Missile casings at The Black Hole surplus store in Los Alamos.

— 44 — 45
Detail of the artwork that became a central feature on the signage above the main entrance of The Black Hole.

— 46

Map of the store created by Tom Grothus for the liquidation sale in 2012.

— 47

Ed in his office at the Shalako Shop, which he ran with Margaret until it closed in 1995. The typewriter was surplus from the laboratory.

— 48 — 49

Letters to Ed Grothus from M. Naraghi of the Atomic Energy Organization of Iran. 1976.

— 50 — 51

Details of different sections of The Black Hole in Los Alamos.

— 52 — 53

Postcard sent to Ed Grothus from Seattle by his son Tom in 1985.

— 54 — 55

Slides from Los Alamos National Laboratory found in The Black Hole.

— 57

Image of an atomic explosion from one of the slides.

— 58 — 59

Protest at Los Alamos against the laboratory's policies.

— 60 — 61

Ed Grothus and Suzuko Terada corresponded regularly for decades, exchanging views about the aftermath of Hiroshima and Nagasaki.

— 62 — 63

Image of an atomic explosion, from the box of slides found in The Black Hole.

— 65

Photograph by Brooklyn photographer Spencer Tunick, taken at The Black Hole, in company with Ed Grothus. December 2000.

— 66 — 67

Photograph taken during President Bill Clinton's visit to Los Alamos. 1998.

— 68 — 69

Ed Grothus printed and distributed over ten thousand postcards in support of his anti-nuclear stance. Some of them were returned to him anonymously, torn into pieces.

— 71

Letter from Ed Grothus to Vice President Al Gore. Ed later received a visit from the Secret Service, after mailing a supposed can of "organic plutonium" to the White House.

— 72 — 73

Ed Grothus at The Black Hole, dressed in his cardinal outfit. Unknown photographer.

— 74

Ed Grothus rhetorically claimed that he celebrated a "critical mass" every Sunday at the First Church of High Technology, located at The Black Hole.

— 76

Ed Grothus inside a large Dewar tank on the grounds of The Black Hole.

— 77

Sketch of the message that Ed Grothus engraved on the monumental twin stone obelisks which he commissioned, but did not live to see erected. The message was translated into fifteen languages. The obelisks were later offered to the town of Los Alamos, but the town refused them.

— 78 — 79

Image from inside Ed's adobe house in Barranca Mesa, Los Alamos.

— 80

The Cerro Grande wild fire in New Mexico. 2000.

The Black Hole

A photo-essay documenting artefacts found during the liquidation sale in 2012.

— 115

View-Master found in Ed Grothus's office.

— 117

Glas-Col heating mantles.

— 118

Object related to optical monitoring/measuring equipment.

— 119

American flags on display at the main counter in the store.

— 120
Decorated bomb casings.

— 122
An early Barber Colman analogue temperature recorder.

— 123
Camera magazines from Los Alamos Scientific Laboratory.

— 125
Graphite bricks used during testing as a shield.

— 126
Motion picture cameras from the Los Alamos Scientific Laboratory.

— 127
A binocular attachment for a microscope.

— 128 — 29
Found books.

— 130
Nuclear Grade HEPA Filter.

— 131
"Organic plutonium – Sparkling water". Artwork by Adam Horowitz, Santa Fe.

— 133
Assorted Dewars flasks, insulated glass bottles for cold liquids storage.

— 135
Interior of a motion-picture projector.

— 136
Free standing health and safety sign.

— 139
Metal container found at the Oppie section.

— 140
Light reflector found in the Laika section.

— 141
Unknown object found at Terra Incognita.

— 142
Solar heat collector cans.

— 145
Part of a gas chromatograph from F&M Scientific Corporation.

— 146
Westinghouse kilowatt hour meters used to measure power consumption.

— 147
Slide projector found in the Oppie section.

— 148
Mechanical calculator.

— 149
Hygrothermograph for recording temperature and relative humidity.

— 151
Telephone handsets found at the Karma section.

— 152
Part of a medical vaporizer used within anesthesia machines.

— 153
Squirrel cage impellers from air moving devices, together with some developing tank spools.

— 154
A larger metal Dewar for storing cold nitrogen.

— 156
Lab container with heating mantle.

— 157
ABE 700 Document perforator, used to mark key documents by punching holes in them.

— 159
Liquid sample found in the Oppie section of the store.

— Cover
Image of the Priscilla Test in 1957, from the box of slides found in The Black Hole.

— Back cover
Ed Grothus in The Black Hole, Los Alamos. Unknown photographer.

ATOMIC—ED

Editorial and Art Direction
Janire Nájera

Archival Materials
Ed Grothus's archive. Los Alamos, New Mexico.

Photographs from The Black Hole
Janire Nájera

Texts
Javier Palao, Celia Jackson & Tom Grothus

Translations
Michal Iwanowski & Janire Nájera

Proofreading
Gregory Dechant

Book Design
Natalia Troitiño

Pre-press
La Troupe

Printed by
Artes Gráficas Palermo

Published by RM

—

ISBN 978-84-17047-61-0
DL: B 25326-2018

CREDITS

RM
© 2019
RM Verlag, S.L.
Loreto 13-15, local B
08029 Barcelona
Spain

© 2019
Editorial RM, S. A. de C. V.
Córdoba 234, int.7, Colonia Roma Norte
06700, Mexico City, Mexico

info@editorialrm.com
www.editorialrm.com

#359

—

VR Viewmaster
Panoramas: Matt Wright
Interaction Design: 4Pi Productions
Graphic design: Maiela Zurutuza

—

This book would have not been possible without the support of Barbara, Tom, Mike Grothus, and Suzi Burns. Special thanks to Matt Wright, who has accompanied me throughout this journey. I would like to extend my sincere gratitude to Natalia Troitiño, Marilyn Phillips, Steve Smith, Marianne Kunz, Richard Krasin, John George, Suzanne Vilmain, Heidi Grothus, Bill Mitchell, Laurie Schrader, Javier Palao, Gareth Phillips, Michal Iwanowski, Charlie Pastor, Lea Tyrallová, Ramón Reverté, Celia Rose Jackson, Elena Mata, Ricardo Nájera, Olga Marín, Jo & Tim Wright, Rich Chitty, Richard Crandon, Chris McFall, Raquel García, Dai Daniel, Ignacio Álvarez, Wales Arts International, and all the generous Kickstarter backers who believed in the project.
www.janirenajera.com

© Janire Nájera 2019. "All Rights Reserved. No part of this book may be reproduced or utilized in any form or by any means, electronic or mechanical, including photocopying, recording, or by any information storage and retrieval system, without express prior permission in writing from the author."

—

© of this edition, RM, 2019
© authors of the photographs
© authors of the texts
© authors of the translations
#onebombistoomany